T0321314

End-to-End
Adaptive Congestion
Control in TCP/IP
Networks

AUTOMATION AND CONTROL ENGINEERING
A Series of Reference Books and Textbooks

Series Editors

FRANK L. LEWIS, Ph.D.,
Fellow IEEE, Fellow IFAC
Professor
Automation and Robotics Research Institute
The University of Texas at Arlington

SHUZHI SAM GE, Ph.D.,
Fellow IEEE
Professor
Interactive Digital Media Institute
The National University of Singapore

Quantitative Process Control Theory, *Weidong Zhang*

Intelligent Diagnosis and Prognosis of Industrial Networked Systems, *Chee Khiang Pang, Frank L Lewis, Tong Heng Lee, Zhao Yang Dong*

Classical Feedback Control: With MATLAB® and Simulink®, Second Edition, *Boris J. Lurie and Paul J. Enright*

Synchronization and Control of Multiagent Systems, *Dong Sun*

Subspace Learning of Neural Networks, *Jian Cheng Lv, Zhang Yi, and Jiliu Zhou*

Reliable Control and Filtering of Linear Systems with Adaptive Mechanisms, *Guang-Hong Yang and Dan Ye*

Reinforcement Learning and Dynamic Programming Using Function Approximators, *Lucian Buşoniu, Robert Babuška, Bart De Schutter, and Damien Ernst*

Modeling and Control of Vibration in Mechanical Systems, *Chunling Du and Lihua Xie*

Analysis and Synthesis of Fuzzy Control Systems: A Model-Based Approach, *Gang Feng*

Lyapunov-Based Control of Robotic Systems, *Aman Behal, Warren Dixon, Darren M. Dawson, and Bin Xian*

System Modeling and Control with Resource-Oriented Petri Nets, *Naiqi Wu and MengChu Zhou*

Sliding Mode Control in Electro-Mechanical Systems, Second Edition, *Vadim Utkin, Jürgen Guldner, and Jingxin Shi*

Optimal Control: Weakly Coupled Systems and Applications, *Zoran Gajić, Myo-Taeg Lim, Dobrila Skatarić, Wu-Chung Su, and Vojislav Kecman*

Intelligent Systems: Modeling, Optimization, and Control, *Yung C. Shin and Chengying Xu*

Optimal and Robust Estimation: With an Introduction to Stochastic Control Theory, Second Edition, *Frank L. Lewis, Lihua Xie, and Dan Popa*

Feedback Control of Dynamic Bipedal Robot Locomotion, *Eric R. Westervelt, Jessy W. Grizzle, Christine Chevallereau, Jun Ho Choi, and Benjamin Morris*

Intelligent Freight Transportation, *edited by Petros A. Ioannou*

Modeling and Control of Complex Systems, *edited by Petros A. Ioannou and Andreas Pitsillides*

Wireless Ad Hoc and Sensor Networks: Protocols, Performance, and Control, *Jagannathan Sarangapani*

Stochastic Hybrid Systems, *edited by Christos G. Cassandras and John Lygeros*

Automation and Control Engineering Series

End-to-End Adaptive Congestion Control in TCP/IP Networks

Christos N. Houmkozlis
Artistotle University of Thessaloniki, Greece

George A. Rovithakis
Artistotle University of Thessaloniki, Greece

CRC Press
Taylor & Francis Group
Boca Raton   London   New York

CRC Press is an imprint of the
Taylor & Francis Group, an **informa** business

CRC Press
Taylor & Francis Group
6000 Broken Sound Parkway NW, Suite 300
Boca Raton, FL 33487-2742

© 2012 by Taylor & Francis Group, LLC
CRC Press is an imprint of Taylor & Francis Group, an Informa business

No claim to original U.S. Government works

Printed in the United States of America on acid-free paper
Version Date: 20120316

International Standard Book Number: 978-1-4398-4057-3 (Hardback)

Library of Congress Cataloging-in-Publication Data

Houmkozlis, Christos N.
    End-to-end adaptive congestion control in TCP/IP networks / Christos N.
Houmkozlis, George A. Rovithakis.
        p. cm. -- (Automation and control engineering)
    Includes bibliographical references and index.
    ISBN 978-1-4398-4057-3 (hardback)
    1. Telecommunication--Traffic--Management. 2. Adaptive routing (Computer network
management) 3. Local area networks (Computer networks)--Traffic. 4. Packet switching
(Data transmission) 5. TCP/IP (Computer network protocol) I. Rovithakis, George A.,
1968- II. Title.

TK5102.985.H68 2012
004.69--dc23   2012005953

Visit the Taylor & Francis Web site at
http://www.taylorandfrancis.com

and the CRC Press Web site at
http://www.crcpress.com

Dedication

Dedicated to my wife Maria, for her support and endurance, as well as my newborn son Nikolaos.
C.N.H.

To my wife Konstantina, and my children, Anastasios and Maria for always being an endless source of love and inspiration.
G.A.R.

Contents

6 Fluid Flow Model Congestion Control   **97**

II   Adaptive Congestion Control Framework   **111**

7 *NNRC*: An Adaptive Congestion Control Framework   **113**

8 *NNRC*: Rate Control Design   **125**

List of Figures

List of Tables

Preface

The Internet is recognized as one of the great achievements that influences implicitly or explicitly the development and the socio-economic prosperity of modern societies. Internet applications like YouTube and Facebook have actually become parts of our culture.

New terms like Internet of Things, Internet of Services and Cloud computing have already appeared. The emergence of diverse applications/services supporting healthcare, education, transportation, administration, trade and entertainment, demands network reliability, availability, interoperability, which in turn impose strict requirements for stability, robustness and fairness, which should be available despite the significant increase of the resulting traffic volume.

Whenever a network link becomes overloaded, congestion appears, leading to packet drops, reduced throughput, increased end-to-end latency that degrades the overall Internet performance. If the phenomenon persists, network congestion collapse may also appear. The need for congestion control is apparent. In fact, the first congestion controller was proposed by Van Jacobson in 1988 just a few years after the appearance of the first congestion collapse. His empirically developed scheme was implemented in the Transmission Control Protocol.

Future Internet requests for congestion control architectures are rigorously designed to operate under extreme heterogeneous, dynamic and time-varying network conditions. The developed controllers should additionally be proven stable and robust against network modeling structural uncertainties and uncontrolled traffic flows acting as external perturbations. Typically congestion control is formulated as an optimization problem trying to allocate, in a fair manner, all network resources among competing sources.

The primary purpose of this book is to establish adaptive control as an alternative though natural framework to rigorously design and analyze Internet congestion controllers. In parallel, a specific adaptive congestion control framework, named *NNRC*, is presented and rigorously analyzed using tools borrowed from adaptive control and approximation theory. Extensions toward cooperation of *NNRC* with application QoS control are also provided.

The book is mainly written for a control audience with the hope of attracting interest in congestion control problems. This is why extensive background material related to computer networks and congestion control is provided. Nevertheless, the book's contents may also be of interest to researchers of Internet congestion willing to investigate adaptive control techniques and concepts. For this purpose, introductory material on dynamic systems stability and neural network approximation is given.

The material related to the *NNRC* adaptive congestion control framework is the outcome of the recent research efforts of the authors. To increase clarity, tables are used to summarize the algorithms that implement the various *NNRC* building blocks. Extensive simulations and comparison tests analyze its behavior and measure its performance through monitoring vital network quality metrics.

Organization of the book

The book is divided in to three parts. A brief summary of each of the parts is given here.

Part I provides background material on computer networks and congestion control and it is comprised of Chapters 2–6. Specifically, in Chapter 2, an introduction to packet-switched networks that represents the controlled system analog of systems theory, is provided. Details related to network connectivity, communication protocols and performance characteristics are given. Chapter 3 discusses how TCP implements congestion control and related issues. Prerequisite to network congestion control is efficient and accurate path congestion measurement. Methods in this direction are presented in Chapter 4. End-to-end congestion control relies on implementing congestion controllers at the source side with minimal information from the network. Representative solutions are given in Chapter 5. Congestion control based on a fluid flow model of the Internet is analyzed in Chapter 6. It is formulated as a resource allocation (optimization) problem. Primal and dual algorithms are derived and major open issues are discussed.

Part II is devoted to the presentation of an adaptive congestion control framework as an alternative to optimization methods. It contains Chapters 7–11. In Chapter 7 the adaptive congestion control framework *NNRC* is overviewed; its main operational modules are presented and qualitatively analyzed. Chapters 8 and 9 are devoted to the rigorous analysis and design of the rate and throughput control modules of *NNRC*, respectively. Illustrative simulations are also provided to highlight the details of the presented algorithms. The performance of the *NNRC* framework is evaluated in Chapter 10 via simulations performed on a heterogeneous long-distance high-speed network, with multiple bottleneck links. Its scalability and dynamic response are also tested under various realistic scenarios. The collaboration of *NNRC* with an application QoS adaptive controller is demonstrated in Chapter 11, and its performance is quantified through simulation studies.

Finally, Part III provides in two appendices background material related to dynamic systems through universal neural network approximators. The results and tools presented in Part III are heavily used in Chapters 7–11.

Using the book

Efforts have been devoted to increasing flexibility and readership through incorporating material related both to computer networks and to adaptive control aspects.

Readers familiar with closed-loop systems and congestion control concepts, and interested only in the application of adaptive control to end-to-end network congestion control, may restrict their attention to Part II (i.e., Chapters 7–11) and probably parts of Appendix B covering function approximation and neural networks.

Control researchers interested in elaborating on the highly complex problem of Internet congestion control may first cover Chapters 1 to 5 to gain the necessary background knowledge on computer networks, before proceeding to the more control theoretic subjects discussed in Chapters 6 to 11.

A computer networks audience interested in incorporating control theoretic techniques to address the congestion control problem may first read Part III to become acquainted with control system concepts and tools and consequently concentrate on Chapters 6 to 11.

Thessaloniki, Greece

Christos N. Houmkozlis
George A. Rovithakis

1

Introduction

CONTENTS

1.1 Overview

It started as an interface experimentation between remote computers in the late 1960s in the U.S. and soon became the most important medium of information exchange, establishing strong connections with the socio-economic fundamentals in most countries worldwide.

The great success of the Internet is driven by

a) its flexibility in networking different types of machinery in terms of hardware architecture, operating systems and communication protocols;

b) its decentralization and self-management.

This lack of a centralized control agency gives an impressive air of freedom that triggered the development of an endless list of applications. YouTube and Facebook have become parts of our culture.

The Internet keeps expanding fast. It has moved from local to national and international. Mobile devices and wireless networks incorporate the Internet Protocol (IP) to establish communication. It is expected that by 2020 the Internet will serve more than 5 billion users around the globe.

The notion of computer networking has been moved towards connecting sensors/actuators and terminals, creating potential for the Internet of Things. On top, diverse applications supporting healthcare, education, administration, trade, comprise the Internet of Services. In addition, the field of entertainment has advanced through 3D videos, interactive environments, network gaming etc.

The emergence of all these applications and services requires a reliable, always available and interoperable network environment, which in turn demands increased levels of robustness, stability and fairness. The high traffic volumes expected can be only partially addressed by over-provisioning.

Excessive load on the network links results in packet loss and reduced throughput. The term used to describe this phenomenon is congestion. Severe congestion leads to network congestion collapse that was first observed in 1986. To address this issue, i.e., to avoid overloading the network resources, Van Jacobson [139] introduced the term congestion control by developing an empirically driven scheme, which was implemented in the Transmission Control Protocol (TCP).

The notion of fairness in congestion control was proposed in [56], visualizing the problem as a fair resource allocation among competing sources. Further generalizations were provided by the work of Kelly et al. [159] who presented congestion control as an optimization problem for networks of arbitrary topology.

The purpose of this chapter is, through the identification of technologies crucial to the Future Internet, to present an overview of at least the dominant new challenges that have to be confronted, and to establish adaptive control as an alternative, though natural framework, through which Internet congestion controllers may be rigorously designed and analyzed.

1.2   Future Internet

Network and service infrastructures are currently constituting the main driving force of development and economic prosperity of the modern societies. On the other hand, mobile communications and broadband Internet have extensively contributed to the substantial growth of the telecommunications sector. Deploying a convergent communication and service infrastructure that will gradually replace the existing Internet, mobile, fixed, satellite and audiovisual networks, is the current research target in the field. It is anticipated that the resulted infrastructure will be pervasive, ubiquitous and highly dynamic, offering unlimited capabilities to users, supporting a wide variety of nomadic and mobile interoperable devices and services.

The increased public awareness of critical shortcomings of the current Internet in terms of performance, security, scalability, reliability as well as economical, societal and business aspects, has led to Future Internet research. The term is probably misleading, as it reflects activities dedicated to the further development of current Internet, not constituting a specific technology.

ITU-T [278] identifies the following technologies as crucial to the Future Internet.

Architecture and Infrastructures
To fulfill the user requirements of flexibility and dependability, current

Internet applications increasingly demand synergies from traditionally separate technologies. Future IP will leverage over simple, to super-fast optical networks. In this direction, the concept of virtualization featuring the construction of optimized virtual networks that meet the demands of a collection of users or applications, appears promising. Currently, a wide variety of wired and wireless access networks have been either deployed or are in the stage of development. Characteristic examples include FTTH/FTTO in the wired domain and 3G/CDMA, WiFi, WiWax, LTE, Satellite in the wireless setting.

The main scope of the Architectures and Infrastructures task is to identify and solve all architectural and infrastructural concerns towards realizing Future Internet. Such issues may include mobility, security, dynamics, awareness of user context, network and service layers, as well as business and socio-economic aspects.

Internet Modeling, Simulation and Measurements

Internet is comprised by the interconnection of highly heterogeneous, complex and dynamic modules. Future developments cannot be accomplished without first comprehending their operational details and behavior. Internet modeling and simulation helps in this direction. In addition, to establish improved network analysis and control protocols, advances in network measurements including Internet probing, network inference and detection of anomalies and attacks are required.

Media Internet

Internet has evolved from file transferring to network gaming, streaming video, telephony etc. In addition, each user is potentially a content provider (peer-to-peer applications). To confront the observed (expected) extreme diversification of the nature and localization of contents, as well as the new interactions with data (semantic Web) and the continuous evolution of applications, the development of new systems, measurements and management tools is required.

Internet of Things

The rapid evolution of pervasive, embedded network devices has reached the point where far more CPUs worldwide are embedded in everyday objects (i.e., cars, domestic appliances, airplanes etc.) than on desktops. The current trend of interconnecting the physical world with sensors and actuators through the Internet will most probably lead to new traffic and architectural challenges. As a consequence, the management of the new devices and network operation to meet the new performance requirements, shall constitute a great challenge.

An interesting example is the extension of the Internet of Things concept to vehicles, with safety being the key potential. Embedded CPUs could communicate road surface conditions to following vehicles, minimizing speed and braking reaction times. Traffic management and on-board pollution sensing can also exploit inter-vehicle and roadside infrastructure networking. Additionally, tracking goods in transit would allow logistic enterprises to optimize their schedules, saving time and energy.

Internet of Services

According to the Internet of Services concept, all resources necessary to use a software application (including the software itself, the tools to develop it and the platform to run it) should be available as a service on the Internet. Cloud computing [208], [240], [260] is the corresponding software terminology used to describe the concept whereby servers, storage, networking, software and information are provided on request. The Internet of Services, when fully deployed, is expected to impact significantly the development of new applications in terms of complexity, time, cost and human effort.

1.3   Internet Congestion Control

Managing Internet traffic and resources to enforce the necessary network characteristics to permit the deployment of the Future Internet is recognized as a crucial task, whose solution (congestion control algorithm) is imperative. In addition, the complexity of the problem increases as any candidate solution should be not only stable but robust as well, against the highly uncertain, dynamic and time varying nature of the Internet.

End-to-end congestion control algorithms have been associated with the Internet transport protocol TCP since the work of Van Jacobson [139] in 1988. Even though they have been proven highly successful, the continuous increment of heterogeneity on both the network and application sides, tend to constrain their performance.

Congestion control incorporates a variety of new challenges that have to be confronted towards Future Internet. In what follows some will be reviewed.

Heterogeneity

Current Internet is a compilation of highly heterogeneous IP networks, realized by different technologies that result in extreme variations in link and consequently in path characteristics. For example, link capacities may range from several Kbps in low speed radio links, to several Gbps in high-speed optical links. The same holds for latency that may start from less than 1ms in local interconnections, going up to seconds in satellite links. As a result, both the available bandwidth and the end-to-end transportation delay may vary significantly, with the trend being rather to increase in the near future. The situation worsens owing to their variation in time, which may even have an abrupt profile owing to dynamic routing, bursty traffic patterns, link changes owing to mobility, topology modifications etc.

The Additive Increase - Multiplicative Decrease (AIMD) congestion control philosophy of TCP assumes a rather static scenario and relatively low bandwidth - delay products. Violation of the aforementioned assumptions leads to low resource utilization.

Congestion control improvements should successfully address the heterogeneity issue, while preserving interoperability and backward compatibility.

Stability

From a control systems perspective, preserving network stability via the appropriate design of a candidate congestion controller is a fundamental task. Typically, stability implies that introducing a bounded input, the network's output should also remain bounded. From the network perspective, such a definition is not necessarily appropriate however, as it does not exclude the appearance of strong oscillations in the link buffers. What is appropriate for congestion control is the notion of asymptotic stability that implies convergence to an equilibrium.

Major prerequisite towards achieving such a goal is a valid mathematical representation of the Internet procedure we want to control. In this way a direct link is established between the control and the controlled variables. Clearly, mathematical modeling of the Internet is a tedious task owing to its inherent complexity, as well as its time-varying and dynamic character. Furthermore, extreme modeling complexity raises significantly the difficulty of designing a controller following formal control-system methods. It is therefore common practice to propose control schemes based on simplifications in both the operating assumptions and in the actual mathematical representation of the Internet, hoping that the produced discrepancies will be taken care of by the robustness inherited when closing the loop.

The verification process is typically performed through extensive simulations on more accurate network models and operating scenarios. For example the impact of the slow-start of TCP is usually neglected in the performance and stability analysis of congestion controllers. Even though this approach is reasonable when analyzing the phenomena appearing at steady-state is of concern, it practically ignores the full extent of the transients. Moreover, it is totally impractical in the study of short-lived http flows that most probably never leave the slow start phase.

Fairness

Besides guaranteeing convergence to a stable equilibrium, candidate congestion controllers should also enforce the fair distribution of network resources among competing sources. It is certainly not a straightforward task that indirectly implies the solution of an optimization problem in the presence of uncertainty in a dynamic and time-varying environment.

In addition to the stated challenges no common fairness metric is broadly accepted by the Internet community. One approach is to consider the approximate equality in the flow rates experiencing equivalent path congestion as will TCP or TCP-friendly sources.

Alternatively, Kelly's work [159] has revealed that as long as a source is accountable for the cost its rate introduces to all other competing sources, then the selected rates are also defined as fair.

Source - Network Cooperation

It is commonly recognized that congestion is an inherent network phenomenon that cannot be resolved without some degree of cooperation between the end-systems and the network. On the other hand, however, such a cooperation should be kept at a minimum level to gain compatibility with the end-to-end operational principle[1] of current Internet, ensuring the scalability and survivability of any new congestion control proposal.

Network components can be involved in congestion control either implicitly or explicitly. In the former, their operation is optimized by properly adjusting the values of a number of free-selected parameters, to support the end-to-end congestion control. In the latter, feedback signals are issued by explicit signaling mechanisms, which are typically realized in the network routers.

Implicit mechanisms are realized via AQM techniques. Even though they are shown to improve network performance in congestion, they suffer from the problem of determining optimal and robust parameterizations. For example many AQM implementations like RED, BLUE, AVQ, PI-Control do not present a systematic way for selecting their parameters.

In explicit signaling, the network exploits two currently unused bits in the packet header to convey information regarding the path congestion status back to the transmitting source, helping the congestion controller to make the necessary decisions towards congestion avoidance.

Robustness

Congestion control should retain high link utilizations and fair resource sharing while preserving robust operation against variations in traffic conditions, and crucial network variables, i.e., link capacities, link buffer size, propagation delays etc.

Network support may help in this direction. However, care should be taken to avoid excessive complexity in the network side. Nevertheless, the congestion controller should be designed in such a way as to avoid compromising performance, through possibly introducing strong oscillations in the network link buffers.

Furthermore, in explicit feedback notification schemes, the mechanism incorporated in measuring congestion and transmitting this information back to the source is by construction erroneous. The two-bits representation of congestion is inaccurate and moreover it does not reflect the current congestion level owing to the presence of possibly large and time-varying delays in the path.

---

[1]Operations that can be realized both in the end-systems and in the network should be actually implemented in the end-systems. The principle ensures that the service provided by the network remains simple, placing any additional complexity at the edges.

Even though significant theoretical progress has been made in the area, the problem in its general description is still open for future investigations.

Non-Congestion Packet Losses

In congestion control, a packet loss is considered as a sign of congestion in the network. Such a conjecture is correct when packets are dropped in routers owing to queue overflows. However, there are other reasons for packet drops. The most typical is owing to corruption. This is mostly the case in networks incorporating wireless links, as well as in UDP where all packets are checked only for corruption and if found corrupted they are rejected.

Apparently, a discrimination unit should be added in the congestion control architecture, to efficiently recognize if a specific packet loss is the outcome of non-congestion mechanisms and consequently not to reduce its sending rate. Explicit feedback notification may help in this direction.

In addition detecting the reason for corruption might be useful towards detecting and isolating faults in the Internet.

Packet Size Effects

Even though the packet size directly affects throughput in TCP, it is not considered in the current Internet. In fact a simplified formula relating the throughput (T_h) of TCP with packets size ($\bar{p}$), end-to-end round trip time (RTT) and packet loss probability (p) is

$$T_h \simeq \frac{\bar{p}}{RTT\sqrt{p}}.$$

Therefore, neglecting the linear dependency of throughput on packet size, high throughput is a result of low RTT and thus of low packet latency. Throughput can be obviously improved by adding an extra control loop on packet size.

Packet size is directly affected when considering Quality-of-Service (QoS) at the application level.

Further research is certainly required along this direction as many open issues remain unsolved. We have discussed for example that the explicit congestion notification mechanism comprises a feedback scheme whose task is to inform the sources about the path congestion level. Should this be done irrespectively of the packet size? Another problem concerns the scalability of congestion controllers with the packet size.

TCP Initialization Effects

During initialization where a connection to a new destination is established, the end-systems have practically no information regarding the path characteristics and the available bandwidth. This lack of knowledge introduces significant uncertainty in deriving a fair rate value.

TCP implements the slow-start mechanism first proposed by Van Jacobson [139]. As it was design based on intuition rather than on solid mathematical arguments, it is far from considered an optimal suggestion in a variety of situations. It may take a long time until the source actually achieves the full utilization of the available bandwidth; the exponential increase policy it follows may be too aggressive, leading to strong oscillations in the queue size of the link buffers, or even worse, to multiple packet loss and large congestion periods. At the same time there is no guarantee of fair behavior.

Slow-start is not incorporated in the stability and performance analysis of at least the majority of the proposed congestion controllers. Its effects are typically demonstrated through extensive simulations.

It is therefore imperative to gain a better theoretical understanding of the initialization mechanism and its impact on congestion, stability and fairness. Furthermore, any new congestion control proposal should be analyzed accordingly, taking all necessary measures at the design phase to minimize the appearance of strong abnormalities.

Misbehavior and Security

Internet congestion control depends for its success on well behaved source-destination pairs. However, both sides are prone to exhibiting a rather selfish behavior; the source intention is to transmit the information obtaining the largest possible percentage of the bandwidth available, while the destination wants to receive the information without necessarily informing the source through feedback about the congestion on the path, as such an obligation costs money. In addition, networks for their own reasons, may desire to make other networks look congested.

The problem becomes increasingly important by noticing that many Internet applications employ transmission protocols that do not use congestion control at all. For example UDP is commonly used in video streaming, which is typically equipped with only a forward error correction procedure to check for packet corruption.

Permitting large volumes of uncontrolled traffic could seriously threaten proper Internet operation, easily leading to congestion collapse especially in low provisioning networks. Malice attacks can be seen as a special case of misbehavior [55], [72].

Possible road maps to addressing these issues include distributed denial of service mechanisms; design of robust congestion controllers against the presence of uncontrolled traffic; enforcing some form of congestion control to UDP applications.

1.4   Adaptive Congestion Control

According to the adaptive control paradigm, information gathered from measuring crucial elements of the system/process to be controlled, are employed to update, in real-time, the controller parameters. Such a tuning procedure, is

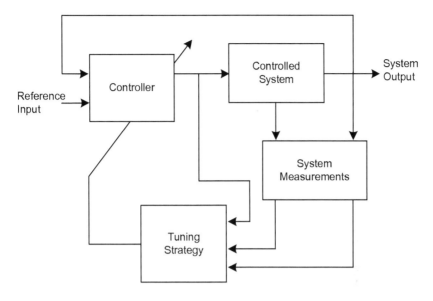

FIGURE 1.1
The adaptive control architecture.

designed to guarantee stability, convergence and robustness of the closed-loop system, despite the presence of significant parametric system uncertainties and exogenous disturbances. The adaptive control architecture is pictured in Figure 1.1.

When the actual controlled system representation is considered uncertain and not just a number of parameters, approximation based control (i.e., neural network control, fuzzy control etc.) can be helpful. Since the early 1990s various research groups [27], [50], [74], [121], [145], [148], [168], [175], [230], [231], [243], [249], [261], [306] have demonstrated the direct link between approximation - based and adaptive control. Roughly speaking, by substituting the actual uncertain system nonlinearities with universal approximation structures such as neural networks, the original problem is transformed into a nonlinear robust adaptive control one.

The discussion in the previous section clearly revealed that regulating congestion can be cast as a standard control problem. Therefore, tools and methods from control systems analysis and design can be borrowed to provide rigorous congestion control solutions. A generalized block diagram of a congestion control scheme is illustrated in Figure 1.2.

A closer look at Figure 1.2 reveals a typical closed-loop control system [95], [102], [181], with the controlled system being the network as seen from the source side. The controller is the congestion control scheme implemented at the source, which receives as input the path congestion status (ECN bit) signaled back through the network by the destination node.

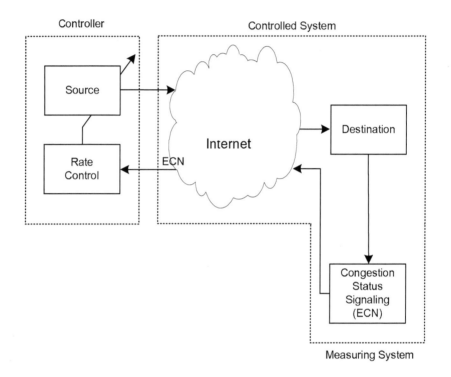

FIGURE 1.2
The Internet congestion control in block diagram form, also showing the analogies with a typical adaptive closed-loop control system.

For the Internet case, the controlled system presents a highly complex, nonlinear, dynamic, time-varying and uncertain process, affected by strong and persistent external disturbances (uncontrolled (UDP) traffic, short-lived flows [14], [165], [307] etc.), as well as inaccurate measurements. The candidate controller should therefore be equipped with the built-in ability to "adapt" fast and accurately, to confront the high degree of uncertainty and complexity in the controlled system (Internet) operation.

In this direction, universal approximation structures can be employed to reduce the level of system uncertainty, providing a valid mathematical representation, which is necessary for the controller development.

If in addition to the above we add the inherent property of tuning in real-time its parameters with minimal human intervention, which has been identified as one of the tasks that deserve further research in congestion control, it becomes apparent that adaptive control comprises a solid framework to rigorously design and analyze congestion controllers for the Internet.

Part I

Background on Computer Networks and Congestion Control

2

Controlled System: The Packet-Switched Network

CONTENTS

2.1   Overview

A computer network is a data communication system which connects two or more autonomous computers [18], enabling data exchange.

Computer networks [218] naturally and gradually emerged from the need for fast interconnectivity utilities as a result of heavy and widely spread computer usage, leading nowadays to concepts like "pervasive" or grid computing. The basic scope of existence of computer networks is the sharing of hardware

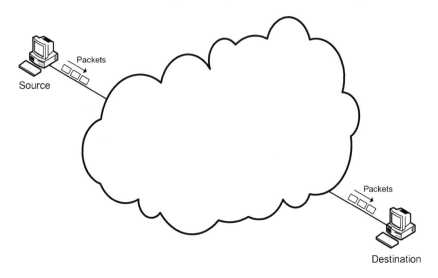

FIGURE 2.1
Packet-switched network.

system resources (memory, storage disks, computing power, printers, etc.) and
the exchange of information in any form (messages, programs, files, data or
even data bases). Those programs, data, devices and resources immediately
become available to anyone upon connecting to the network, irrespective of
their actual physical position. As a consequence, these characteristics lead to
money saving, resource economy, increase of system utilization, central control
and system scalability. The largest known computer network is the Internet
[62], mainly comprised of packet-switched sub-networks.

Packet-switched networks [169] are the backbone of the data communication infrastructure. A simple packet-switched network is shown in Figure
2.1. A source breaks down every piece of information, such as an e-mail message, into small chunks called packets.[1] Packets are routed to their destination
through the packet-switched network. The destination computer reassembles
the packets into their appropriate sequence, thus reconstructing the initial
piece of information. Packet switching is employed to optimize the use of the
bandwidth available in a network and to minimize the latency, a measure of
time delay experienced in a system.

Remark 2.1 *Packet-switched networks constitute an improvement over the
circuit-switched networks, which formed the basis of the conventional telephone
systems and were the only existing personal communication infrastructures
prior to the invention of packet-switched networks. In conventional telephone
networks, a circuit between two users must be established for a communication*

---

[1]In some circumstances, packets are also required to be attached together or further
fragmented, forming a new packet known as a frame.

to occur. Circuit-switched networks required resources to be reserved for each pair of end users. This implies that no other user could use the already dedicated resources for the duration of network use. The reservation of network resources for each user resulted in an inefficient use of bandwidth especially for applications in which information transfer was bursty.

In the rest of this chapter issues related to network topology and its operation will be discussed. In addition, several performance characteristics will be analyzed and their relation to well-known applications will be reported.

2.2 Network Connectivity

Towards network connectivity among a set of computers, several secure and stable connections should be established in such a way that user privacy is protected and network scalability is not affected.

Even though some networks operate by limiting the set of machines that are eventually connected, others (with Internet being the prime example) are designed to grow in a most scalable way that potentially allows them to provide interconnections to all the computers in the world, creating a well established "Web Graph" [35], [252] that consists of nodes and links.

2.2.1 Links and Nodes

In an abstract way of networking, computers are the nodes, directly connected by physical mediums, the links. Several physical mediums for such links exist, including copper wire, fiber optics, microwaves, infrared and communication satellites.

Two types of connection links are illustrated in Figure 2.2. In the first, the physical links are limited to a pair of nodes. Such a link is said to be point-to-point. In the second type, more than two nodes may share a single physical link. Such a link is said to be broadcast (or multi-link). Whether a given link supports point-to-point or multiple-access connectivity depends on how the node is attached to the link. It is also the case that broadcast links are often limited in size, in terms of both the geographical distance they cover and the number of nodes they connect. The exception is the satellite link, which can cover a wide geographic area.

2.2.2 Sub-Networks

As an illustrative example for a sub-network, let us suppose that we want to interconnect 6 remote nodes. If we are restricted to direct connectivity, an

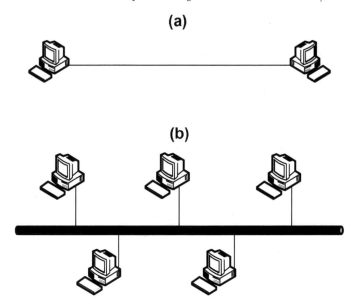

FIGURE 2.2
Types of connection links: a) point-to-point link; b) broadcast link.

all-to-all solution appears (see Figure 2.3a), with obvious disadvantages such as the increased cost and administrative complexity.

The alternative scheme, depicted in Figure 2.3b, is the packet switched network developed with the help of indirect links. Such indirect links are implemented with the cooperation of intermediate nodes, called switches [254]. Thus, in this concept, connectivity between two nodes does not necessarily imply a direct physical connection between them, and indirect connectivity may be achieved among a set of cooperating nodes.

In Figure 2.3b the two intermediate nodes play the role of the switch, and the link that connects them is designated as the shared link. Two links are attached to each switch whose main function is to forward data to the appropriate output link.

The cloud in Figure 2.3b encapsulates the network switches, the nodes inside it, and is surrounded by the hosts, the nodes outside the cloud that use the network. In this way numerous packet-switched sub-networks can be synthesized to form a large scale network consisting of many such clouds. Throughout this book we shall use the cloud icon to denote any type of network, whether it is a single point-to-point link, a multiple-access link, or a switched network.

In Figure 2.4, a set of independent networks (clouds) is interconnected to form an inter-network, or internet for short. The node that is connected to two or more networks is commonly called a router, and it plays much the same role as a switch-it forward packet from one network to another. Appar-

(a)

(b)

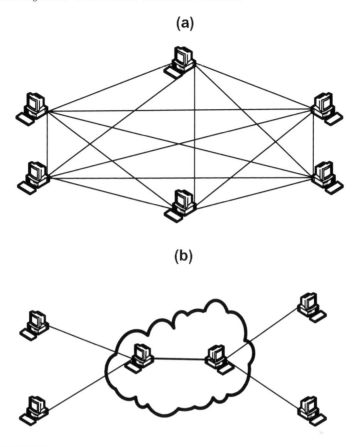

FIGURE 2.3
a) The network tries to connect 6 nodes using only direct connections; b) a packet switched network for the same 6 nodes.

ently, arbitrarily large networks may be constructed by simply interconnecting clouds.

2.2.3 Network Classification

Computer networks are classified [271] by their range. If they range a few meters, they are classified as personal area networks (PANs). Networks that range a few hundred meters are classified as local area networks (LANs). When grouping several local area networks together, within a range of some kilometers, the network is classified as a metropolitan area network (MAN). Any network ranging more than some kilometers is classified as a wide area network (WAN). LANs and WANs constitute the two main categories of computer networks.

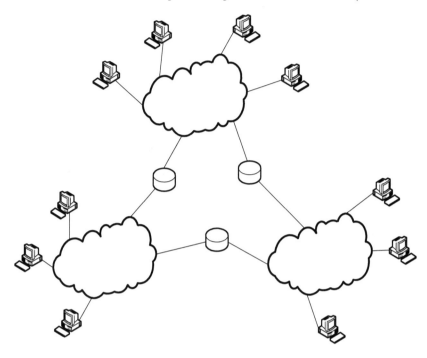

FIGURE 2.4
Many interconnected clouds form an inter-network, or internet.

A LAN is extended in a geographically limited region (e.g., a building or group of buildings). They are widely used to connect personal computers and workstations in company offices and factories, to share resources and exchange information. A WAN spans a large geographical area, often a country or a continent. It contains a collection of machines intended for running user (i.e., application) programs. LANs can be connected via special hardware to form either a bigger LAN or, if the distance is long, a WAN.

Generally, LANs are distinguished from WANs by three characteristics:

1. size,

2. transmission technology,

3. topology.

As LANs are restricted in size, smaller delay of data transmission is experienced. Knowledge of this delay makes it possible to use certain designs that further simplify network management.

LANs typically use multi-access transmission technology, which means that a single communication channel is shared by all computers in the network. Traditionally LANs run at speeds of 10 Mbps to 100 Mbps (even though speeds up to 10Gbps have also been reported), introduce low delays (microseconds or nanoseconds) and make very few errors.

2.2.4 LAN Topologies

Topology is the network architecture used to interconnect the networking equipment.

Computer networks are designed by purpose and importance. While some networks require high bandwidth and high reliability, others require high bandwidth and low cost. In what follows the most common network topologies are summarized.

Line Topology

FIGURE 2.5
Line topology.

In a line topology, each node is connected to at most two neighboring nodes. Data transmitted from one end of the network to the other end will have to travel through all network nodes [112], [263]. Figure 2.5 is illustrative.

This network topology is easy to create, and can span large distances as each node may act as a repeater. However, as it lacks redundancy, its reliability is highly dependent on the performance of all other nodes in the network. In addition, if one node fails, this will cut the network in two, as there are no alternative routes [112], [263]. The dominated advantages and disadvantages are summarized in Table 2.1.

Bus Topology

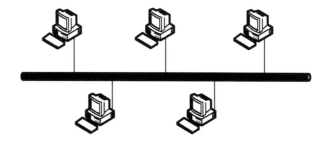

FIGURE 2.6
Bus topology.

Bus networks (see Figure 2.6) use a common backbone to connect all nodes. A node wanting to communicate with another network node sends a packet

TABLE 2.1

Advantages and disadvantages of network topologies

Network Topology	Advantages	Disadvantages
Line	a) Implementation b) Long-distance	a) No redundancy b) Reliability dependent on the performance of each network node
Bus	a) Implementation b) Management	a) Small-size networks
Ring	a) Management	a) Low reliability b) Performance sensitive to node modifications
Star	a) Simple operation b) Node isolation	a) Reliability highly dependent on the operation of the central hub
Tree	a) Long-distance b) Reliability with respect to peripheral node failures	a) More cabling b) Very sensitive to central hub failures

onto the wire that all other nodes see, but only the intended recipient actually accepts and processes the packet [112], [263].

If a node fails, network reliability is not affected, but if a link fails, communication is lost as there is no alternative route [112], [263]. Furthermore, the bus topology is easy to handle and implement. However, it performs well only for a limited number of nodes, thus restricting its use to small size networks.

Table 2.1 summarizes the main advantages and disadvantages of the bus topology.

Ring Topology

In a ring topology, each node is connected to two other nodes, forming in this way a ring (see Figure 2.7). The ring topology is often the most expensive, and it tends to be inefficient, as information travels through more nodes, compared to other topologies [112], [263]. Each node has equal communication access.

Typically, in a ring topology information flows either clockwise or anti-clockwise. Such a restriction impacts reliability as a node failure (or a cable break) results in network failure. Moreover, modifications in any network node directly affect the performance of the entire network. In fact, it can be visualized as a line topology with all its weaknesses. A way to solve this problem is to use a dual ring topology, where each node has four branches connected to it. This makes the topology more resistant to failures, but it also increases

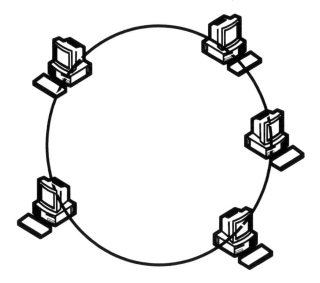

FIGURE 2.7
Ring topology.

cost [112], [263]. With respect to network management, no central server is required, reducing cost and effort.

In Table 2.1 the basic advantages and disadvantages of the ring topology are reported.

Star Topology

In the star topology, each node is connected to a central node, which either retransmits all data received to all the other nodes connected to it,[2] or processes the incoming data which are consequently retransmitted only to the destination node [112], [263]. Figure 2.8 is a graphical representation of the star network topology.

If a connection between a node and the central node is broken, this will only lead to the isolation of that node from the rest of the network. But if the central node is broken, the entire network will fail [112], [263]. This is why highly reliable hardware is used for the central node. Moreover, network redundancy is achieved with the help of spanning tree algorithms [112].

Within local area networks, this is the most common network topology, as it requires the least amount of transmission medium, and allows for network adaptivity.

The advantages and disadvantages of the star topology are underlined in Table 2.1.

---

[2]In this case, although the arrangement is physically a star, it operates like a bus.

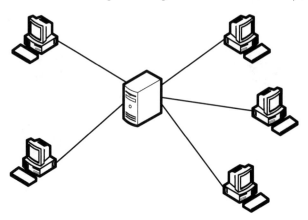

FIGURE 2.8
Star topology.

Tree Topology

In a tree network topology the nodes are arranged as a tree, (see Figure 2.9) and operate exactly as they would if they had been connected to a ring or to a bus topology. The nodes connected as non-leaves also act as they would if they had been connected to a ring or to a bus topology, but they also have several network cards for possible connection to other leaves.

It is important to note that no routing is performed at the non-leaf nodes; they simply relay data from their input to their output, like any other node [112], [263].

If a link to a leaf or a node itself fails, it will only result in the isolation of that leaf node. The rest of the network will still be operational. However, if a non-leaf node fails, an entire section of the network will collapse [112], [263].

In addition, using a central hub as a repeater increases the distance a signal can travel between nodes.

The main advantages and disadvantages of the tree topology are provided in Table 2.1.

2.3   Network Communication

2.3.1   Packet Switching

In a packet switching scheme, prior to transmitting a piece of information, the source divides the total amount of data into fundamental units of information called packets. Each packet is individually transmitted through the

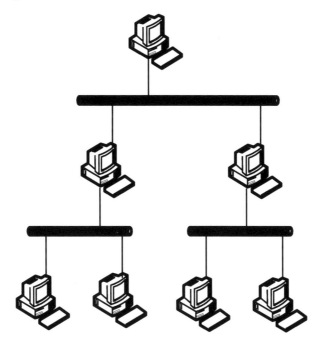

FIGURE 2.9
Tree topology.

network. The destination computer re-assembles the packets appropriately to reconstruct the initial piece of information.

Depending on the network, each packet can be of fixed or of variable size and consists mainly of two parts: the header and the body. The body contains a segment of the data being transmitted.

The header contains a set of instructions regarding the packet's data, including the sender's address, the intended receiver's address, the number of packets into which the data has been divided, the identification number of the particular packet, the packet length[3] and the synchronization.

Packets are switched through networks via routers located at various network points. Routers are specialized computers that forward packets through the best paths, determined by the routing algorithm [206] being used on the network, to the destinations indicated by the destination address located in the packet header. All packets of a source do not necessarily follow the same path to reach their destination. Therefore, they can be received in a different order (out of order phenomenon).

Packet-switched network operation typically uses a strategy called store-and-forward. As the name suggests, each node in a store-and-forward network first receives a packet over some link, then stores the packet in its internal

---

[3]Applies to networks that support variable packet length size.

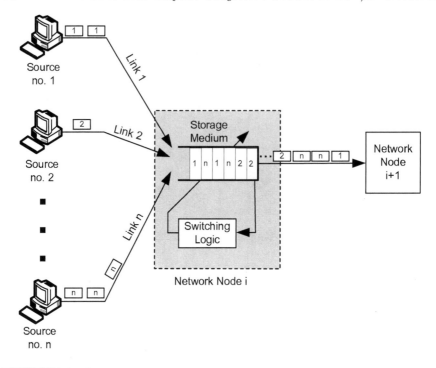

FIGURE 2.10
Illustration of the store and forward strategy. The arrow traversing the storage medium denotes the network switch which, according to a switching logic, reorders the already stored packets that consequently are placed onto the single output link.

memory, and consequently forwards the packet to the next node. Store and forward is implemented through network switches, which are special devices performing packet multiplexing. Figure 2.10 is illustrative.

2.3.2   Protocols and Layering

To efficiently exchange data between two hosts, a single communication link is insufficient. Concurrently, both the source and the destination are forced to follow some typical abstract rules and materialize concrete processes. For example the source must first enable a direct communication link to the destination via the network infrastructure, which undertakes the task to restore such a link. Furthermore, the source should know when the destination is ready to receive data. Also, every host must be able to know if the transmitted data have been correctly received by the destination. Thus, communication between computers on a network requires the use of a set of rules, forming the communication protocol [122].

Generally, protocols for computer networking use packet switching techniques to send and receive messages. Some network protocols include mechanisms for devices to identify and make connections with each other, as well as formatting rules that specify how data is packaged into the sending/receiving messages. Other network protocols also support message acknowledgement and data compression and are designed for reliable and/or high-performance network communication. Nowadays, a surprisingly large number (of the order of hundreds) of different computer networking protocols have been developed, each designed for specific purposes and environments.

To reduce the design complexity, most networks are organized hierarchically as a stack of layers called levels, where each layer is built upon the one below it. The number of layers, the name of each layer, as well as its contents and functionalities, differ from network to network. The purpose of each layer is to offer certain services to the higher level layers, shielding those layers from the underlined details of how the offered services are actually implemented. In a sense, each layer is a kind of virtual machine, offering certain services to the layer above it.

In this hierarchical organization, instead of creating a single protocol to implement all communication functionalities, a number of protocols assigning each specific set to its corresponding layered structure is developed. The role the levels play, the relationship between them and the protocols to be followed all determine the concept of the network architecture.

OSI Architecture

The typical and most commonly used seven-layer network architecture is illustrated in Figure 2.11 and called the OSI architecture [296].

Each level performs a small subset of functions required for communication. It builds its functionality on the previous lower level, which performs the primary functions, and provides its services to the next higher level. Ideally, all the levels would be defined in such a way that changes in one level do not require changes at other levels. Thus, the problem of communication is divided into smaller and more manageable tasks.

The complete seven layered OSI model covers various network operations, equipment and protocols. The lower level is closer to the underline hardware and the higher to the application. Each layer communicates with the one immediately above and below it. The lower three layers are implemented on all network nodes, including routers within the network and hosts connected along the exterior of the network.

In Table 2.2 operational details concerning the seven layered OSI architecture are provided. It can be noticed that both the layers and their defined tasks ensure an optimum separability map for all parties involved in the networking industry.

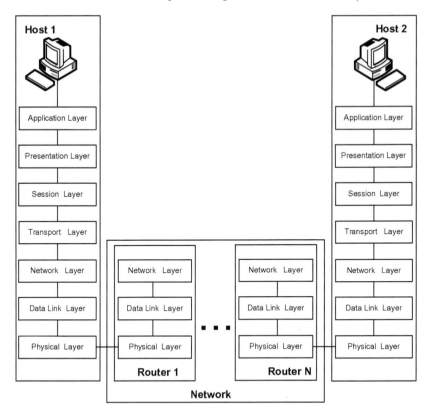

FIGURE 2.11
OSI architecture.

2.3.3　Internet Architecture

The well known Internet is a "huge" collection of cooperating computer networks. The Internet was designed and developed to ensure reliable communication between the several disparate and heterogeneous systems that compose it. The set of communication protocols that implement the Internet is the TCP/IP architecture [192] also called Internet architecture.

The adoption of TCP/IP architecture is not in conflict with the OSI model, as the two systems were developed simultaneously. However, there are some fundamental differences between the two models. As in the OSI case, TCP/IP is also a layered architecture. However, the layers of TCP/IP are not in a one-to-one correspondence with the layers of OSI. Their actual relationship is clearly illustrated in Figure 2.12.

Complete correlation between levels of the two models can only be observed for the transport and network layers. The application, session and presentation layers of OSI correspond to the application layer of TCP/IP, while the data

TABLE 2.2

Operational details of the OSI architecture

Application Layer	Supports application and end-user processes.
Presentation Layer	Provides independence from differences in data representation by translating from application to network format, and vice versa. The presentation layer works to transform data into the form that the application layer can accept.
Session Layer	Establishes, manages and terminates connections between applications. The session layer sets up, coordinates and terminates conversations, exchanges and dialogues between the applications at each end.
Transport Layer	Provides transparent transfer of data between end systems and is responsible for end-to-end error recovery and flow control. It ensures complete data transfer.
Network Layer	Provides switching and routing technologies, creating logical paths for transmitting data from node to node. Addressing, internetworking, error handling, congestion control and packet sequencing are functions of this layer.
Data Link Layer	Ensures the reliable transfer of information on physical line connector. Transmits the frames and handles errors in the physical layer, flow control and frame synchronization. The data link layer is divided into two sub layers: The Media Access Control (MAC) layer and the Logical Link Control (LLC) layer. The MAC controls how a computer on the network gains access to the data and permission to transmit it. The LLC controls frame synchronization, flow control and error checking.
Physical Layer	Deals with issues of physical cabling and bit physical transmission. Defines electrical and physical specifications for devices. In particular, it defines the relationship between a device and a transmission medium.

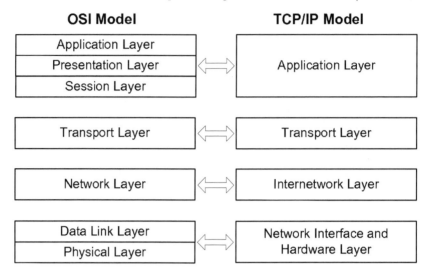

FIGURE 2.12
The relationship between the layered architecture of OSI and TCP/IP Model.

link and physical layers of OSI, to the network interface and hardware layer
of TCP/IP.

Details of the TCP/IP architecture are shown in Figure 2.13, in which
the communication protocols that implement the functions of each level are
specifically mentioned.

Application	FTP, TFTP, SMTP, http, RTP, etc		
Transport	TCP	UDP	
Internetwork	IP		ICMP
			ARP/RARP
Network Interface and Hardware	Ethernet, Token Ring, FDDI, PPP, etc.		

FIGURE 2.13
The stack of TCP/IP model protocols.

In what follows, a brief presentation of the basic functionalities of each
level of the TCP/IP model will be provided.

Network Interface and Hardware Layer

At the lowest level of the TCP/IP architecture, which it is called Level Access Network, lie those communication protocols whose principal function is to transmit packets among nodes of the computer network. The nodes communicate either with a point to point link or through a broadband link. A wide variety of protocols have been developed and widely used for transmitting packets over a link (e.g., Ethernet, Token Ring, FDDI, PPP, etc.). Usually, these protocols are implemented with a collaborative hardware component (e.g., a network adapter) and a piece of software (e.g., the corresponding network adapter driver).

Internetwork Layer

At the network layer of TCP/IP the dominant protocol is IP (Internet Protocol), with the main responsibility of controlling the addressing of the network and the routing of packets. The IP, however, cannot guarantee the delivery of all packets and in the right order to their destination. The IP protocol runs over all possible network infrastructures to make it possible to link many different network technologies and combine them into one logical Internet.

Internet access may be established either from home, through the traditional social telephone network (PSTN), or from a local network at the work place, or even when traveling, via wireless telecommunication systems (e.g., the mobile network). All these different Internet access technologies have as a common characteristic the use of IP.

Transport Layer

At the transport layer of TCP/IP we find the protocols of TCP (Transmission Control Protocol) and UDP (User Datagram Protocol), which both control the exchange of packets between the hosts, setting the end to end communication. The UDP protocol delivers a packet to its destination by performing only a simple check to see if the packet has been altered in transit through the network. If worn, then it is rejected, otherwise the packet is forwarded for further processing. Instead, TCP performs a more complex error checking. Specifically, if a packet is found damaged, retransmission is requested. In addition, TCP mechanisms allow packet flow control that reduces the transfer rate in the presence of a network congestion situation until normalized. Thus, TCP/IP networks have important functions of error control and flow control, carried out at the hosts of the network, relieving in this way the intermediate nodes from performing complex and costly operations.

Application Layer

Thus far, it is described how the data stream is split into datagrams that consequently are being transferred from one host to another and how, ultimately, it is reconstructed at the destination host. However, to enable communication at the application level, we need something more. For example, to transfer a file which is located on a remote computer, we first establish a connection and gain access to this remote computer, then the file is requested and transmission is started. All these actions are implemented by the application protocols at the application layer, which run over TCP/IP.

When the TCP protocol opens a connection, the application protocols visualize the connection as a simple wire that connects the source to the destination. Application protocols determine what to send through this link. To allow different types of computers to communicate, each application protocol defines the presentation of data. Note, however, that both the TCP and IP protocols are not interested in the presentation of data.

There is a wide variety of application protocols such as FTP (File Transfer Protocol), TFTP (Trivial File Transfer Protocol), SMTP (Simple Mail Transfer Protocol), HTTP (Hyper Text Transfer Protocol), TELNET (protocol to access remote computer), RTP (Real - Time Transfer Protocol), SNMP (Simple Network Management Protocol), DNS (Domain Name), NFS (Network File System), etc. Their main function is to ensure the interoperability of the respective applications. To understand the difference between the application protocol and the implementation of the corresponding application, consider for example the commercially available software tools used for reading web pages (e.g., Netscape Communicator, MS-Internet Explorer, the Mosaic etc.). All these applications comply with the rules of application protocol HTTP. The consequence is that we can use any browser to access an electronic web site on the Internet. Most application protocols use the services of TCP to communicate with their peer protocols (e.g., FTP, SMTP, HTTP, TELNET, RTP). Needless to say, there exist application protocols that use UDP (e.g., TFTP, SNMP, NFS), as well as some that use both (e.g., DNS).

2.3.4   Transfer Control Protocol (TCP)

The main workhorse in transport layer protocols of TCP/IP technology is the Transmission Control Protocol (TCP). TCP was constructed to provide connection-oriented services, reliable data transfer and end to end communication between hosts in packet-switched computer networks (the latter is schematically shown in Figure 2.14).

The transport layer (which runs TCP) always receives continuous data streams from the application level. According to TCP, every continuous data stream must be segmented into several packets if it exceeds a predefined threshold. Consequently, these packets are transmitted through the internet system, enabling the final communication between the two hosts. The threshold used to define the packets to be sent is agreed upon the establishment of a TCP connection. Each of these packets defines the typical transport unit in the TCP protocol, which is called a segment. A segment consists of a header and the body of the transmitting data. The header has several fields created by the TCP sender and used to help every TCP receiver to properly manage the receiving segments. The format of a TCP packet is shown in Figure 2.15.

The definition of the segment order is based on the corresponding field

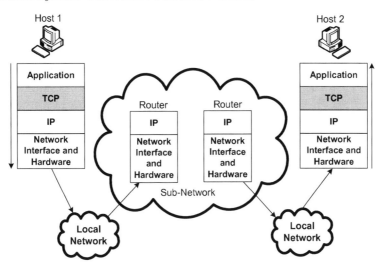

FIGURE 2.14
End-to-end communication.

of the header, called sequence number. If for example, the sequence number field is set to 3, this means that the specific segment is the third of the series defined when the original data were split. Moreover, the sequence numbers are also used at the destination to eliminate duplicates.

To ensure that a segment has reached its destination, the latter must send back an acknowledgment (ACK), which is a TCP segment without data, having thus only a header. More specifically, when acknowledging, the destination places a number in the field of the segment header indicating that all data until that byte number have been received correctly. The corresponding field is called ACK number. For example, if a TCP receiver sends a segment with ACK number equal to 1500, it means that the destination has correctly received all the first 1500 bytes of data. If the source does not receive an ACK within a reasonable time frame, it retransmits the data segment. This time interval is called timeout and the process, retransmission.

While the two aforementioned fields in the segment's header are used to perform important tasks specifically oriented for each of the two communicating sides, the "window" field is a major for both the TCP source and the destination. To further elaborate on this, we must first analyze another primary function performed by the TCP protocol, that of controlling the amount of data that can be transmitted each time. This feature is known as flow control. It is not reasonable for a TCP sender to wait for an ACK confirmation to send the next segment. This will result in an unwarranted reduction in the rate of transmission through enlarging the idle time. On the other hand, the destination limits the acceptance rate of the data received. For example, if the upload speed is much greater than the rate of absorption of the destination,

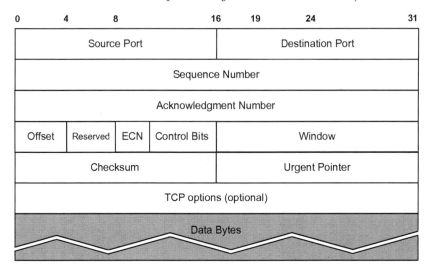

FIGURE 2.15
TCP segment details.

the area for temporary storage, i.e., the buffer, is quickly overwhelmed, leading to a forced rejection owing to buffer overflow, potentially causing a major waste of network bandwidth. Therefore, it is necessary for both ends of each connection to indicate how much new data they can accept. That information is kept at the "window" field of the segment's header.

Furthermore, the TCP header contains 6 control bits (TCP flags), which are used to relay control information between TCP hosts. Possible TCP flags include:

- URG: signifies that this segment contains urgent data.

- ACK: is set any time the Acknowledgment field is valid.

- PSH: signifies that the source invoked push operation.

- RST: signifies that the destination has become confused.

- SYN: used when establishing a TCP connection.

- FIN: used when terminating a TCP connection.

Finally, the Checksum field monitors the header, the data and the pseudo header (see Figure 2.16). The pseudo header contains the Source Address, the Destination Address, the Protocol, and TCP length and protects TCP against misrouted segments. This information is carried in the Internet Protocol and is transferred across the TCP Network interface.

Every host has several ports, each one dedicated to a specific network

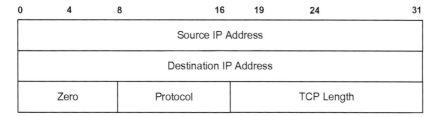

FIGURE 2.16
The pseudo header of a TCP segment.

application (telnet, ftp, http etc.), as well as a unique network address (IP Address). Both the source and the destination port fields are stored in the two corresponding fields of the segment's header. The source port combined with the source IP address forms the so-called source socket. In the same manner, the destination port concatenated with the destination IP address forms the destination socket. This pair of sockets uniquely identifies a TCP connection. In this way a socket may be simultaneously used in multiple connections.

All mechanisms concerning the reliability and flow control described above require that TCP initialize and maintain a certain status of information for each data stream. The combination of this information, that regards sockets, sequence numbers, and window sizes, is called a TCP connection. When two applications wish to communicate, TCP must establish a connection (initialize the status information on each side). When their communication is completed, the connection is terminated to free the system resources for other uses.

TCP Connection Establishment

To establish or terminate a connection, TCP follows a specific procedure according to which both the source and the destination have to state their starting sequence numbers, information crucial to establish a robust communication. This is achieved by exchanging three messages (after which the algorithm took its name), the three-way handshake (see Figure 2.17a for an illustration).

The procedure is initiated by the source transmitting the initial sequence number. The destination acknowledges the received sequence number and informs the source about his own. The procedure terminates with the source issuing a second message actually acknowledging the correct arrival of the destination's sequence number.

As can be noticed from Figure 2.17, both parties acknowledge the actual sequence number plus 1, implicitly identifying the next sequence number expected. Even though it is not clearly shown, a timer is also scheduled and if the expected response is not received in time, then retransmission is authorized.

The TCP connection is terminated when the source transmits a FIN packet, and the appropriate acknowledgement issued by the destination is correctly received by the source. Figure 2.17(b) is illustrative.

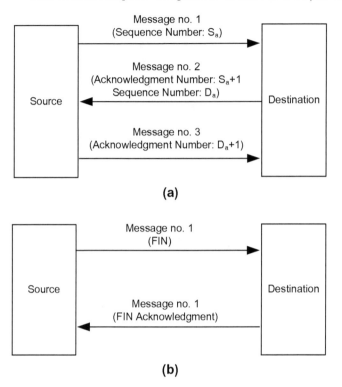

FIGURE 2.17
Illustration of the three-way handshake algorithm: a) initialization phase; b) termination phase.

Packet Retransmission

One of the most important procedures in TCP is the way it handles the retransmission of packets that are lost. Each time a packet is sent, the TCP source initiates a retransmission timer, waiting for acknowledgment. If the time set for the timer expires prior to receiving the packet acknowledgment, the TCP source concludes that the packet is lost, or that it has arrived damaged at its destination, and retransmits the packet.

TCP is designed to operate in an Internet environment. A TCP packet while traveling through the Internet may have to traverse many sub-networks of different philosophy, transmission speed and size, connected through multiple routers. As a result, the source cannot determine in advance the time required for a packet to reach the destination and receive an ACK, denoted as the round trip time (RTT). Another factor affecting the RTT of a packet is the status of the sub-networks through which it passes, which is constantly changing over time.

For all these reasons, TCP is designed to modify its behavior in the presence of different network delays, using an adaptive retransmission algorithm.

Essentially, TCP examines the performance of each connection and produces the appropriate values for the retransmission timers. The performance of the connection changes as the TCP recalculates the values of the timers.

To collect all data needed by the adaptive retransmission algorithm, TCP records the time instant a packet was sent and the time instant it receives its ACK. Using those two time instants, the TCP produces the RTT. Every time TCP receives a new value for the RTT, it recalculates the weighted average of RTT (denoted as $\overline{RTT}$), through the formula [62]:

$$\overline{RTT}(new) = \alpha\overline{RTT}(old) + (1 - \alpha)RTT \qquad (2.1)$$

where the parameter α, $(0 < \alpha < 1)$ controls the effect on $\overline{RTT}$ of the RTT. If a value close to 0 is chosen, then $\overline{RTT}$ would react to RTT variations significantly faster than selecting a value of α close to 1.

When sending a packet, TCP calculates the value of the retransmission timer as a function of the current value of $\overline{RTT}$ according to:

$$Timeout = \beta\overline{RTT} \qquad (2.2)$$

where β is a constant parameter $(\beta > 1)$. The timeout must be always higher than the average round trip time. If $\beta = 1$, then a small increase in delay over $\overline{RTT}$ would initiate packet retransmission, magnifying unnecessarily the network traffic. On the other hand, a large β increases the retransmission tolerance of TCP, with negative effects on the total transmission delay.

2.3.5 User Datagram Protocol (UDP)

UDP is a communication protocol that offers a limited amount of service when data are exchanged between hosts in a network that uses the Internet Protocol (IP). UDP is an alternative to TCP and, together with IP, is sometimes referred to as UDP/IP. Like TCP, UDP uses the Internet Protocol to transmit a data unit (called a datagram) from one host to another. However, unlike TCP, UDP does not provide the service of dividing a large data packet into small packets (segments) and of reassembling at the other end. Therefore, the application program that uses UDP must be able to guarantee that the entire data packet has properly arrived.

The format of the UDP segment is shown in Figure 2.18. The segment starts with the source port, followed by the destination port. These port numbers are used to identify the ports of applications at the source or the destination, respectively. The source port identifies the application that is sending the data. The destination port helps UDP to demultiplex the packet and directs it to the right application. The UDP length field indicates the length of the UDP segment, including both the header and the data. UDP checksum specifies the computed checksum when transmitting the packet. If no checksum is computed, this field contains all zeroes. When this segment is received at the destination, the checksum is computed; if there is an error, the packet is discarded.

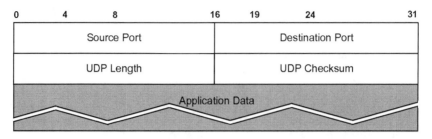

FIGURE 2.18
UDP segment details.

UDP is used mainly in applications where the critical factor is speed. Voice is a typical application example. In voice applications, packet retransmission does not make sense, as the principal aim is to minimize the delay introduced by the protocol, whose presence deteriorates the received speech quality.

2.3.6   Internet Protocol (IP)

The Internet Protocol (IP) constitutes the main network layer protocol of TCP/IP technology. Its operation is based on the idea of datagrams (a basic transfer unit associated with a packet-switched network), which are transferred independently, from source to destination, without ensuring "reliable" service. An "unreliable" service does not notify the user if delivery fails. All controls for reliable transmission of data are performed at the transport layer and more specifically at TCP.

Each time a TCP (or a UDP) protocol wants to transmit a segment, it forwards the data to the IP protocol, specifying also the address of the destination host. This address is the only element that is of interest to the IP. Thus, IP does not care about segment content or how it is related to previous or next segments. Each time the IP receives a TCP (or a UDP) segment, it adds its own headings forming in this way an IP datagram, whose total length is 64 Kbytes. From the moment the IP protocol forms the IP datagram, its role is limited to tracking down the proper path to the destination. The format of an IP datagram is shown in Figure 2.19.

Having the path determined, the data segment is transmitted through physical networks, which follow the rules of the Network Interface and Hardware Layer of the TCP/IP model. The physical networks, depending on the technology followed, may use a maximum length of the transport unit different from the 64 Kbytes of the IP datagram. To address such an issue, the IP protocol breaks down the datagrams into smaller parts, called fragments. These fragments, are reconstructed when they reach their final destination to form the original datagram. Such a decomposition takes place at the first router of the path, which has to forward the datagram through a physical network with maximum packet length less than the length of the datagram. The fragments

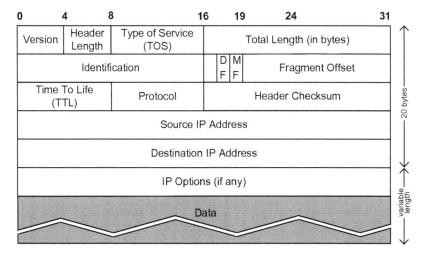

FIGURE 2.19
IP header fields.

created constitute new datagrams, and follow possibly independent paths. In order for the destination IP protocol to determine which datagram each received fragment belongs to, it uses the Identification field of the IP header. All fragments of the same datagram have identical values in this field.

To distinguish at the destination if an incoming packet belongs to an original datagram or to a fragmented one, the More Fragment (MF) field is used. If the MF field is set to 1, it means that the datagram is broken into several fragments. If for some reason the destination host is unable to assemble the datagram, it sets the Don't Fragment (DF) field at 1. If the datagram has to be fragmented to be routed (it must be forwarded to a network that cannot handle the datagram without breaking it up into smaller pieces), the router will disregard it and will send an error message back to the originating host.

To determine the order of the fragment in the original datagram, the Fragment Offset is utilized.

Apart from the fields of the IP header, which already have been mentioned, very important fields for the IP operation are the Internet addresses of the source and the destination, the Protocol Number and the Checksum. The source address determines the IP address of the sending host (the field is necessary so that the opposite host knows who sent the datagram). The destination address is the IP address of the host to which the datagram must be delivered. The Protocol Number field informs the IP protocol of the opposite host, to which higher level protocol must deliver the datagram (e.g., TCP or UDP). The Protocol Number takes a value corresponding to the transport layer protocol. The Checksum field allows the IP protocol of the opposite side to verify the correctness of the datagram header. Such a control is necessary because the datagram header is constantly modified (at each router) as it travels the network, increasing in this way the error probability.

The Version field is used to determine the version of IP protocol, which owns the datagram. It is important that everyone involved in managing a datagram follow the same protocol version.

The Header Length field indicates the length of the header in words of 32-bits and specifically it determines the fixed part of the header (Figure 2.19). The IP Options field is used for special functions of the protocol. The Total Length field gives the length of the whole datagram (header and data). The maximum length is 65.536 bytes. In cases where a datagram is broken into fragments, the field gives the length of the specific fragment and not of the entire original datagram.

The Type of Service field is used to inform the host regarding the kind of service request offered by the communication network. The characteristics that define the service offered by the network and used by IP are: throughput, reliability and delay.

The Lifetime field (TTL) is used by the host to determine the life time of datagrams. Each time a datagram passes through a router, the field is reduced by one. When the field takes the zero value, the datagram is rejected. This field is used to destroy datagrams that have either lost their path and thus have been overdelayed to reach their destination, or an error has occurred in the destination address.

2.4 Performance Characteristics

The behavior and performance of TCP/IP protocols is evaluated by measuring and calculating various performance metrics [79] (queue size, bottleneck link utilization, throughput, packet loss rate, delay, fairness). This section introduces several performance indicators, that shall be used extensively throughout the book.

2.4.1 Queue Size

Queue size indicates the number of pending packets in the router's queue. One of the main objectives of TCP/IP protocols is to stabilize the queue size. An unstable system often leads to queue oscillations and strong synchronization among TCP flows. In that respect, router queue size is an important performance characteristic of TCP/IP networks.

2.4.2 Throughput

Throughput is defined as the data rate at which a source can send packets (including retransmitted packets) to a destination. In other words, if a source successfully delivers N packets of constant size p_{size} to a destination at a specific time window T_w, the resulted throughput T_h is

$$T_h = \frac{Np_{size}}{T_w}.$$

2.4.3 Link Utilization

The link utilization (U) measure, reveals the utilization of the bottleneck links in a network. Good utilization of the bottleneck link results in improved network traffic behavior. For a link with maximum capacity c_l, its utilization is calculated via

$$U(\%) = \frac{T_h p_{size}}{c_l} 100$$

where T_h is the throughput and p_{size} is the size of the transmitted packets.

2.4.4 Packet Loss Rate

Packet loss rate indicates the number of packets dropped, owing to router buffer overflow, and is mostly used for congestion notification purposes. If N_d packets are dropped out of N packets transmitted, then the packet loss rate L is calculated by

$$L = \frac{N_d}{N} 100.$$

Increased packet losses result in retransmissions, which in turn lead to reduced network efficiency.

2.4.5 Round Trip Time

Round trip time (RTT) [4], [106], [149], [157], also called round trip delay, is the time required for a packet to travel from a source to its corresponding destination and back again. On the Internet, an end user can determine the RTT from the IP address by pinging that address. Ping is a computer network administration utility used to test the reachability of a host on an Internet Protocol (IP) network and to measure the round trip time for messages sent from the originating host, to the destination computer. The RTT incorporates both propagation and queueing delay and depends on various factors including:

- The source data transfer rate.

- The nature of the transmission medium (copper, optical fiber, wireless or satellite).

- The physical distance between the source and the destination.

- The number of nodes in the path connecting the source and the destination.

- The amount of traffic on the LAN to which the end user is connected.

- The number of other requests being handled by intermediate nodes and the remote server.

- The speed at which all intermediate nodes and the remote server operate.

- The presence of interference in the path.

In a WAN or the Internet, RTT is one of the most significant factors affecting latency, which is the time between the request for data and the complete return or display of that data. The RTT can range from a few milliseconds (under ideal conditions between closely spaced points) to several seconds under adverse conditions between points separated by a large distance.

The theoretical minimum of RTT is the propagation delay. In satellite communications systems this minimum time can be considerable, since the RF signals may have to propagate tens of thousands of kilometers through space.

2.4.6 Fairness

Fairness has been defined in a number of different ways. Currently, there's still no generally-agreed-upon definition for fairness in a computer network. However, different measures of fairness have quite different implications. Fairness can be considered between flows of the same protocol and between flows using different protocols. Fairness can also be considered between sessions, between users or between other entities.

Jain's Fairness Index

Jain's fairness index [294] postulates that the network is a multi-user system, and derives the metric to see how fairly each user is treated. It's a function of the variability of throughput across each user. For a set of users throughput $(x_1, x_2, \ldots, x_n)$, Jain's fairness index is defined as follows:

$$J(x_1, x_2, \ldots, x_n) = \frac{1}{n} \frac{(\sum_i x_i)^2}{\sum_i x_i^2}. \tag{2.3}$$

The Jain's fairness index always lies between 0 and 1. A value of 1 indicates that all flows got exactly the same throughput. This index equals k/n when k users equally share all network resources, while the remaining n-k users receive zero allocation.

Max-Min Fairness

Max-Min fairness [234] is a common fairness definition which is also called bottleneck optimality criterion. A formal definition follows [30], [144].

A set of throughputs $\{x_i\}$, $i = 1, 2, ...n$ is said to be max-min fair if for any other set of throughputs $\{y_i\}$, $i = 1, 2, ...n$ that satisfy the network capacity constraints it holds: if for some $i \in \{1, 2, ..., n\}$ $y_i > x_i$ then $\exists k \in \{1, 2, ..., n\}$ such that $y_k \leq x_k \leq x_i$.

Clearly, under max-min fair allocation, to increase the $i - th$ source's throughput (x_i), we have to decrease the throughput allocated to some other source, whose throughput is less than or equal to x_i.

It can be proven [262] that a set of throughputs is max-min fair if and only if every source has a bottleneck link. In other words, all max-min throughput allocation algorithms unavoidably result in bottleneck links.

Proportional Fairness

A feasible allocation, x_1, x_2, ..., x_n, is defined as weighted proportionally fair [159], if for any other feasible allocation x_1^*, x_2^*, ..., x_n^*, the weighted sum of the proportional changes in each user's throughput is less than or equal to zero. If all a_i's are equal to one, then it is simply called proportionally fair. In other words:

$$\sum a_i \frac{x_i^* - x_i}{x_i} \leq 0. \tag{2.4}$$

This criterion favors smaller flows, but less emphatically than max-min fairness.

2.5 Applications

In this section we shall present a number of broad Internet applications beginning from traditional to more recent ones. For the latter the dominant technological issues that restrict their performance are clearly underlined. All Internet applications are classified with respect to their architecture and to their demands in performance characteristics.

According to the architecture, applications are divided into peer-to-peer [22], [220], [73], [207], [188] and client-server [29], [60], [46], [190], [236], [237] (see Figure 2.20). In client-server applications, clients communicate with each other through a server, which acts as a middleman, keeping the master copy of all information, running nearly all application logic and downloading the results to the client. In peer-to-peer applications, however, virtually all application logic and information reside to the client, which communicates directly with other clients without server intermediation. Although some peer-to-peer applications are hybrids, with server components, the server isn't the dominant player; it's merely a referee, standing off on the sidelines, overseeing the game without playing it.

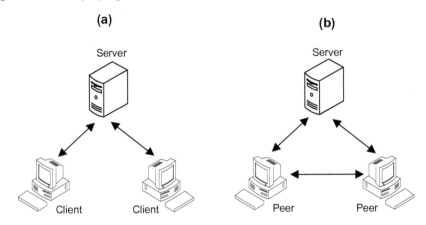

(a)   **(b)**

FIGURE 2.20
a) Client server model diagram and b) peer-to-peer model diagram.

According to the performance requirements, applications are classified into real time and off-line. Usually, real-time applications require high bandwidth and small delays in data transfer. Reliability is not the main issue. The exact opposite is usually the case with off-line applications.

2.5.1   E-Mail

In an e-mail application, users exchange digital messages across a computer network. E-mail systems are based on a store-and-forward model, according to which servers accept, forward, deliver and store messages. Users are no longer required to be online simultaneously and need only connect briefly, typically to an e-mail server, for as long as it takes to send or receive messages. An e-mail message consists of two components, the message header, and the message body, which is the e-mail's content. The message header contains control information, including, minimally, the originator's e-mail address and one or more recipient addresses. Usually additional information is added, such as a subject header field.

In TCP/IP technology, to transport an e-mail, the Simple Mail Transfer Protocol (SMTP) is used, which is based on a client-server model. E-mail protocols have no quality requirements except reliability.

2.5.2   World Wide Web

Word Wide Web (WWW) [49], [212], [241] is an information dissemination system located in geographically dispersed areas. This system combines technical revocation of information with the technology of hypertext, creating an easy to use global information system. Web information is structured in the form of hypermedia, including text, pictures, sound files, video and generally any kind of multimedia. In other words, the Web is a graphical way of representing and transmitting information. This environment allows activation of different links, which lead to information, wherever they are located on the Internet. The separate sections of hypertext, which are highlighted on the computer screen, are called Web pages. References of one section to another are called links. The reading of a hypertext is realized with tools, called Web browsers.

The Protocol used to transfer hypertext is the Hypertext Transfer Protocol (HTTP), which is based on a client-server model. The Web servers are systems permanently connected to the Internet, which host the websites that are available for access. Browsers (Web clients) are used to access the servers and read Web pages.

HTTP requires reliable and fast data transfer of web pages, especially if these pages contain multimedia data. Clearly, such requirements can be fulfilled by conventional TCP/IP protocols.

2.5.3 Remote Access

Remote access applications allow a user working on a computer to connect with a remote computer and execute programs on the latter through the terminal. The commands typed at the terminal are transferred through the connection to the remote computer and the user works as if he were in front of the remote computer. The remote computer can be on the same local network, or another network of the same company, or anywhere on the Internet. The only requirement is that the user has permission to access the remote system.

The protocol that implements the remote access is Telnet and follows the client-server model. Telnet requires reliable and secure transfer of its data. Clearly, such requirements can be handled by conventional TCP/IP protocols.

2.5.4 File Transfer

The File Transfer Protocol (FTP) [185] allows transferring files between computers, using TCP/IP technology. FTP is based on the client-server architecture, and uses the reliable end to end service provided by the transport layer protocol TCP. Specifically, FTP allows the creation of a file copy from one system to another. Thus, a user, who works on a computer, can efficiently send or receive files to another computer. The system security is ensured by the implementation of control authorization for each user requesting access to the system. The authorization check is carried out using a username and password, which are assigned by the system administrator and are checked each time a user requests access to the system. Under FTP, a user has full access to the system, but permission only to copy files. When the connection to the remote system is restored, FTP allows users to copy one or more files on their own computer. The term transfer indicates that the file is transferred from one system to another, but the original file is not affected. FTP priority is the reliable data transfer.

Additionally, in the case of transferring large files, adequate bandwidth is also required to achieve a sufficiently small transportation time. This application can also be handled by conventional TCP/IP protocols.

2.5.5   Streaming Media

Streaming media [255], [13] is a technology that enables the inclusion of audio, video and other multimedia elements that visitors to a website will be able to listen to or view immediately, without having to download the file to their own computer. Streaming media typically consist of audio only, video with audio or any combination of audio, illustrated audio, video, synchronized graphics or animation. More specifically, live streaming means taking the video and broadcasting it live over the Internet. One of the most popular forms of live streaming is the P2PTV.

P2PTV refers to peer-to-peer (P2P) software applications designed to re-distribute video streams in real time on a P2P network. The distributed video streams are typically TV channels from all over the world. The main drawback of P2PTV is that it is bandwidth demanding. Hence, its growth is limited by congestion problems. Besides congestion, fairness is another restrictive mechanism. In a non-fair operation, some P2PTV connections may consume more bandwidth compared to others, limiting their performance. Currently, P2PTV operates over the UDP protocol, which does not support either congestion or fairness control. Clearly, protocols engaging such control modes in their operation will impact positively the future of P2PTV.

2.5.6   Internet Telephony (VOIP)

Voice over Internet Protocol (VoIP) [191] is simply the transmission of voice traffic over IP-based networks. VoIP has become popular mainly owing to its reduced cost over traditional telephone networks. Many companies and individuals are taking advantage of the development of new technologies that support the convergence of voice and data over their existing packet networks (Figure 2.21). This has created a new role for the network administrator, that of a telecommunications manager. The network administrator must not only be aware of the issues of data transport within and external to the network, but also the issues of incorporating voice data traffic into the network. An example of VoIP application is Skype.

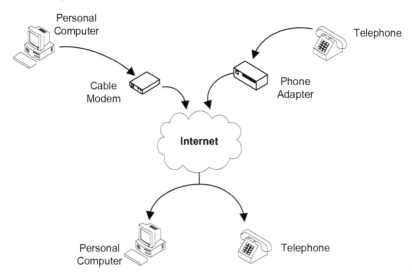

FIGURE 2.21
Voice over IP.

As the number of users increases, the total number of connections rises even more rapidly, resulting in high bandwidth demands and consequently to congestion problems. Clearly, such issues affect negatively the requests for improved Quality of Service and reliability of VoIP applications.

2.6 Concluding Comments

Internet is a network which has no centralized governance in either technological implementation or policies for access and usage. Each user tries to satisfy its own requirement for throughput and Internet protocols try to satisfy the requirements of all users. When the user requirements for throughput remains at low levels, there is no problem in protocols to achieve their goal. However, the extensive use of real time multimedia applications in communication and business services, such P2PTV and VoIP, imposes the requirement for throughput increment. The work of internet protocols is even more difficult along with the heterogeneity of the networks, especially considering the increased usage of wireless networks, as well as the tendency of internet service providers to offer users control over the behavior of applications, known as Quality-of-Service (QoS) [41], [42], [68], [77], [100], [105], [111] management. The necessity of satisfactory quality service delivery has introduced QoS as a fundamental part of various protocols and mechanisms that enable computing and communication systems.

Owing to their time-varying character, it is very hard to predict the exact multimedia application requirements (i.e., bit rate in video). To achieve quality assurance, multimedia services have to be easily adapted to resource availability. Application adaptation is accomplished by adapting application characteristics according to the network availability. Many applications can adapt to dynamic network conditions because they either have built-in adaptation mechanisms or may collaborate with internet protocols.

The possibility of cooperation between applications and protocols gives great possibilities to adapt the application characteristics to the available throughput. For example, in a teleconference application, if the throughput capability of the network is degraded, then the performance of the video communication may be degraded as well (lower resolution, slower frame rate, less color depth). However, at the same time, the most important services of the teleconference application, such as the voice and data transfer connection, should be maintained in high quality (clear sound, high-speed transfer of documents). In the next chapter we shall discuss in detail the operation of TCP protocol weaknesses in synchronous networks.

3

Congestion Issues and TCP

CONTENTS

3.1   Overview

Current Internet is a best effort network, as it does not provide any guarantee in data delivery and the users obtain unspecified variable bit rate depending on the traffic load. Internet applications share on equal terms all network resources (e.g., link bandwidth). Whenever the aggregate demand for a link bandwidth exceeds the available link capacity, congestion occurs. As a reaction to congestion, the router which feeds the specific link typically places all excess packets in a buffer, which roughly operates as a basic FIFO queue [10], and drops packets only if the queue is full. Resulting effects include long delays in data delivery, wasted resources, and even possible congestion collapse, in which all communication in the entire network ceases.

Static solutions, such as allocating more buffer space or providing faster links, prove insufficient for congestion control purposes. For instance, an increase in the buffer capacity may enlarge packet delays. On the other hand, an increase in link speed increases the possibility of congestion, owing to the apparent mismatch at the point of interconnection of a high-speed with a low-speed network. Furthermore, as we have seen in Chapter 2, video and voice applications require smooth operating sources, in addition to congestion avoidance. It is therefore clear that in order to maintain network performance above a sufficient level, certain mechanisms must be provided to prevent the network from being congested, or if unavoidable, to guarantee that the congested time period is minimized.

This chapter focuses on the presentation and analysis of the algorithms and mechanisms utilized by current TCP, to manage network congestion. Their drawbacks are also underlined, establishing the necessity of developing more advanced methodologies to overcome their limitations.

3.2   Core Issues in Congestion Control

As already mentioned, a network is considered congested when the number of packets that are forwarded for temporary storage at the buffer of an intermediate router exceeds its capacity, resulting in an amount of packets being dropped, and forcing the network to collapse. To avoid such an unpleasant situation, a congestion control mechanism is mandatory. A well designed congestion control scheme must operate both proactively (when congestion is expected) and reactively (while congestion is detected). In addition, congestion control should provide sufficient services under heavy load to avoid adverse effects on the network operation, as for example source transmission rate oscillations.

To achieve both reactive and proactive operations, a congestion control mechanism consists of a source algorithm that dynamically adjusts source rates based on congestion in their paths, and a router algorithm that updates, implicitly or explicitly, a certain congestion measure at each router link and feeds it back, implicitly or explicitly, to all sources that use the link.

Besides congestion collapse, fairness is of major concern in best effort networks. The issue of fairness among competing flows has become increasingly important for several reasons. The most significant one is to avoid the situation where a few flow streams occupy most of the bandwidth, acting abusively towards other flows. The type of fairness achieved is strongly related to the congestion control algorithm utilized by the sources.

A third reason for a flow to use congestion control is to optimize its own performance regarding throughput, delay and loss rate. In certain circumstances, as for example in environments of high statistical multiplexing, the delay and loss rate experienced by a flow are largely independent of its own sending rate. However, in environments with lower levels of statistical multiplexing or with per-flow scheduling, the delay and loss rate experienced by a flow are in part a function of its sending rate. Thus, a flow can use congestion control to limit the delay and the losses experienced.

In environments like the current best-effort Internet, fairness and congestion collapse concerns limit the range of congestion control behaviors available to a flow.

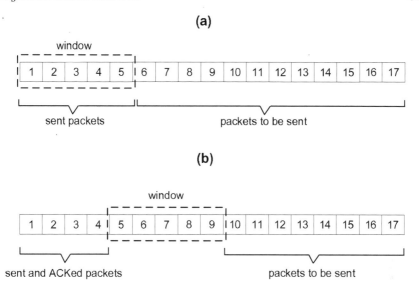

FIGURE 3.1

The "sliding window" concept. The destination defines the window size and the source starts transmission. For every acknowledged packet, the window slides forward.

3.3 TCP: Flow Control and Congestion Control

Current internet is dominated by the presence of TCP, which is a connection-oriented unicast protocol that offers reliable data transfer, as well as flow and congestion control. In this section we will present how TCP regulates packet flow and will analyze its performance.

One of TCP's primary functions is to properly match source transmission rate to that of the destination and the network. A 16-bit window field is used by the destination to communicate to the source the amount of data bytes it is willing to accept. As the length of the window field is limited to 16 bits, a maximum window size of 65535 bytes is resulted.

This information is called advertised window (rwnd). As packets are transmitted and appropriately acknowledged, the window slides forward to cover more data in the byte stream. This concept is known as the "sliding window" and is depicted in Figure 3.1.

Packets within the window boundary are eligible to be transmitted by the source. Packets behind the window have already been sent and acknowledged. However, packets ahead of the window have not been sent and must wait for the window to "slide" forward before they can be transmitted by the source. The window size is adjusted at the destination each time an acknowledgment is issued. The maximum transmission rate is ultimately bounded by the destination's ability to accept and process data. The aforementioned technique implies an implicit trust arrangement between the TCP source and the destination.

TCP flow control is called to successfully address two controversial tasks. On one hand, high transmission rates should be guaranteed, while on the other, the network should be protected against the congestion effects. The mechanism employed by current TCP to manage congestion is comprised by four algorithms namely:

1. Slow start

2. Congestion avoidance

3. Fast retransmit

4. Fast recovery

These algorithms were devised in [139] and [141] and their use within TCP was standardized in [38].

3.3.1   Slow Start

The slow start algorithm is used to start packet transmission and to detect the available bandwidth. The latter is necessary to avoid the appearance of congestion to a network (through transmitting large bursts of data) whose operating condition is initially unknown to the source.

To implement slow start, two extra variables are utilized. The congestion window (cwnd) and the slow start threshold (ssthresh). The first sets an upper limit on the amount of data a source can transmit over the network before receiving an acknowledgment (ACK). The second variable (ssthresh) is used to determine the activation or deactivation of slow start.

When a new connection is established, the congestion window is initialized to one segment[1] (a segment is any TCP data or acknowledgment packet), and waits for its ACK. When that ACK is received, the congestion window is incremented from one to two, allowing the transmission of two segments. In fact the source may transmit at most $min\{cwnd, rwnd\}$ packets. The difference between those windows is that cwnd is imposed by the source, while rwnd is imposed by the destination. The former is based on the source assessment of perceived network congestion; the latter is related to the amount of available buffer space at the destination side.

Continuing the slow start presentation, when each of those two segments is acknowledged, the congestion window is increased to four. This provides an exponential growth; although in reality it may not be exactly exponential because the receiver may delay its ACKs, typically by sending one ACK for every two segments it receives. In this way, excessive traffic is avoided. Schemati-

---

[1]Experimental TCP variations exist [6], [71] that permit larger initial windows.

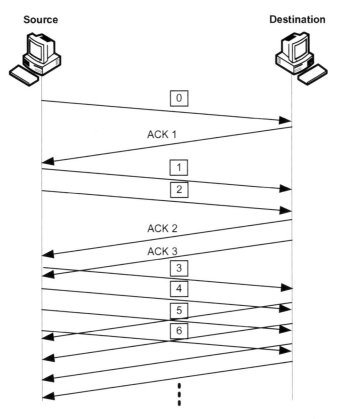

FIGURE 3.2
A schematic description of the slow start algorithm.

cally the slow start algorithm is described in Figure 3.2 and the exponential increase of cwnd is illustrated in Figure 3.4.

Slow start operates until

$$cwnd < ssthresh.$$

Besides the aforementioned criterion, slow start also terminates whenever congestion is experienced.

Initially ssthresh may be chosen arbitrarily high (for example, in some implementations ssthresh is selected equal to rwnd), but it may be reduced in response to congestion.

3.3.2 Congestion Avoidance

After slow start, the TCP flow control enters the congestion avoidance phase and remains within until congestion is detected. The congestion avoidance algorithm assumes that packet losses owing to non-congestion issues are very small (significantly less than 1%). Therefore, the loss of a packet signals congestion somewhere in the network between the source and the destination. One formula commonly used to update cwnd during congestion avoidance is:

$$cwnd(new) = cwnd(old) + \frac{1}{cwnd(old)}. \tag{3.1}$$

This adjustment is executed on every incoming non-duplicate ACK. This is a linear growth of cwnd, compared to the slow start's exponential growth. The increase in cwnd should be at most one segment for each round trip time (regardless of how many ACKs are received in that RTT), whereas in slow start cwnd is incremented by the number of ACKs received in a round trip time. Schematically, the congestion avoidance algorithm is described in Figure 3.3, while the linear increase of cwnd is depicted in Figure 3.4.

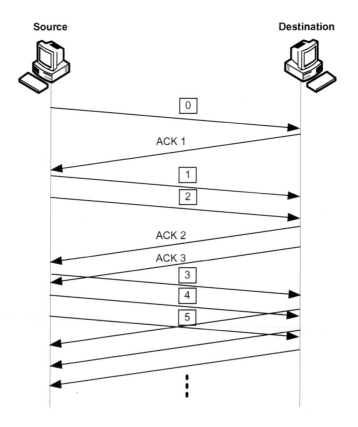

FIGURE 3.3
A schematic description of the congestion avoidance algorithm.

Upon detection of a segment loss, the TCP source sets ssthresh to:

$$ssthresh = max \left\{ \frac{FlightSize}{2}, 2 \right\} \qquad (3.2)$$

where FlightSize corresponds to the amount of outstanding packets in the network.

Congestion avoidance and slow start are independent algorithms with different objectives. Whenever congestion occurs, TCP reduces its transmission rate and invokes slow start to re-initialize transmission. Figure 3.4 is a visual description of both the slow start and the congestion avoidance algorithms.

Further details on the topic may be founded in [7], [139], [264].

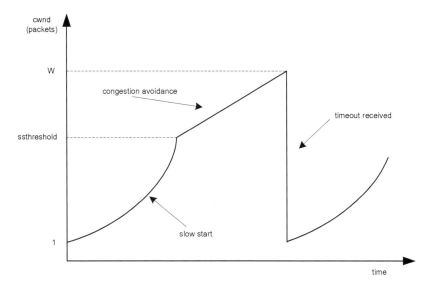

FIGURE 3.4
CWND variation within slow start and congestion avoidance algorithms.

3.3.3 Fast Retransmit and Fast Recovery

When a segment arrives at its destination out of order, a duplicate ACK (DU-PACK) is immediately issued, whose purpose is to inform the corresponding source accordingly and additionally to indicate which sequence number is expected.

In a network, duplicate ACKs may be issued for the following reasons [294]:

- Owing to dropped segments. In such a case, all segments after the dropped one will trigger duplicate ACKs.

- Owing to re-ordering of data segments.

- Owing to replication of ACK or data segments.

As TCP does not distinguish if a duplicate ACK is caused by a lost segment or is just a reordering of segments [20], [28], [180], it waits for a small number of duplicate ACKs to be received. It is assumed that if there is just a reordering of segments, there will be only one or two duplicate ACKs before the reordered segment is processed, which will then generate a new ACK. Receiving three or more duplicate ACKs in a row is a strong indication that a segment has been lost. TCP then performs a retransmission of what appears to be the missing segment, without waiting for the retransmission timer to expire. The above described process is called fast retransmit and was first defined in [141].

After fast retransmit sends what appears to be the missing segment, congestion avoidance and not slow start is performed. This is the fast recovery algorithm which governs the transmission of new data until a non-duplicate ACK arrives. It constitutes an improvement towards high throughput under moderate congestion, especially for large windows. The reason for not performing slow start in this case arises from the fact that the destination generates the duplicate ACK only when another segment is received. Therefore, as there is still data flowing between the two ends, it is not wise for TCP to reduce the flow abruptly by going into slow start. Figure 3.5 is a visual description of fast retransmit and fast recovery.

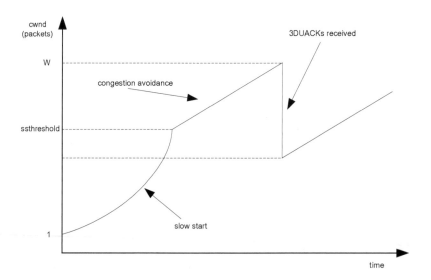

FIGURE 3.5
Visualization of fast retransmit and fast recovery algorithms.

Further details concerning the implementation of both processes can be found in [82], [83].

3.4 TCP Problems

In current TCP, packet losses are utilized to estimate the available bandwidth in the network. As long as no packet losses are experienced, TCP continues to increase its window size by one packet per round trip time. In the presence of a packet loss, the window size is reduced to half. This is the so called Additive Increase / Multiplicative Decrease (AIMD) policy. This abrupt behavior is the main source of problems which are briefly discussed below.

Ambiguity Loss
As already mentioned, TCP considers segment loss as an indication of congestion along the communication path. However, a packet may be lost for reasons not directly connected to congestion. For example, wireless links are known to experience sporadic and usually temporary losses owing to fading, shadowing, hand off and other radio effects that cannot be considered congestion. In such cases, TCP incorrectly identifies the packet losses as congestion effects and drastically decreases the window size, causing the underutilization of the radio link.

Start-Up Behavior
To find the appropriate sending rate, each TCP connection probes constantly the network for available bandwidth using the slow start algorithm. However, slow start is a time consuming process, especially when high-speed long-distance networks are considered. Actually, it may take a significant number of RTTs in slow start, before the TCP connection begins to fully utilize the available bandwidth of the network. For instance, assuming an initial congestion window of 4 segments, a total of $log_2(N) - 2$ round trip times are required [294] before $cwnd = N$. Practically, this means that the slow start phase in a long-distance high-speed network could easily exceed 5 seconds.

Furthermore, the exponential growth of cwnd at slow start may result in a vast amount of packet losses in networks with high bandwidth-delay products. The phenomenon, which is also called slow start overshoot, results in unnecessarily large timeout retransmissions, eventually leading the TCP connection to enter the congestion avoidance phase with a very small congestion window. Recovering from such a situation typically requires a significant number of RTTs.

RTT Unfairness
During the congestion avoidance phase, TCP increases its congestion window by one packet in a round trip time. Thus, when flows with different round trip times compete for a single bottleneck link, in the average, flows with lower round trip times will obtain more bandwidth from the bottleneck link than flows with high RTT, as they increase their congestion window less often.

Reordering

Packet reordering is the situation in which Internet packets from the same flow are being received in a different order than the one in which they were sent [216].

As already discussed, segments received out of order cause immediate duplicate ACKs to be issued by the receiver. The arrival of 3 consecutive duplicate ACKs triggers the fast retransmit/fast recovery algorithm at the source which is spurious in this case, degrading performance/link utilization. Numerous studies [176], [225], [228], [147], [134], [28] have shown that packet reordering is common, especially in a high-speed networks where there is a high degree of parallelism and different link speeds. Possible reasons for packet reordering include:

- Link layer recovery.

- Handovers.

- Priority blocking.

- Route changes.

- Multipath routing.

- Differential services.

Delay Spikes

As long as the round trip time varies within the boundaries of the timeout, TCP operates normally. If a delay spike does exceed the timeout, TCP will retransmit the oldest outstanding segment and will enter slow start with cwnd set to one packet. The aforementioned behavior causes deterioration for the following reasons: Firstly, starting-over with slow start causes poor link utilization. Secondly, the spurious retransmissions will trigger out of order duplicate ACKs to be sent by the receiver, triggering fast retransmit/recovery at the source, which in turn will degrade performance even further. Typical reasons for causing delay spikes include [110], [108], [135]:

- Link layer recovery.

- Handovers.

- Priority blocking.

- Temporal link outages.

- Route changes.

Scalability

TCP does not scale well with the growth of bandwidth modern link technologies provide. The problem stems from the fact that the size of the congestion window oscillates between the bandwidth delay product and one segment. As the bandwidth delay product increases, the oscillation time also increases. For usual gigabit links, this time is at the magnitude of one hour.

Designed to Be Congested

TCP is designed to fill router queues and even to make them overflow. Filling queues increases latency and overflowing queues wastes bandwidth. This attitude leads to an oscillation in the round trip delay of the packets. This oscillation results in larger delay jitter and low utilization of the available bandwidth owing to the retransmissions issued after each packet drop.

3.5 Managing Congestion

The TCP problems described in the previous section, dictate the necessity of developing new congestion control protocols. However, all possible new suggestions should be subject to several constraints. In this section the most important are summarized.

3.5.1 TCP Friendliness

A novel congestion control protocol should operate in a fair manner, converging rapidly to variations in network conditions. Additionally, it should be TCP-compatible, meaning it should be capable of ensuring that the TCP connections using AIMD get their fair allocation of bandwidth in the presence of these protocols and vise versa. One notion that has been proposed to capture TCP-compatibility is "TCP-friendliness."

TCP-friendliness is measured via the effect on throughput of a non-TCP flow on the competing TCP flows, under the same conditions. A non-TCP unicast flow is said to be TCP-friendly, if it does not affect the long term throughput of any of the co-existing TCP flows by a factor that exceeds the one achieved by a TCP flow under the same conditions.

Throughput in a TCP connection mainly depends on round trip time (RTT), retransmission timeout value (RTO), segment size (SMSS) and packet loss rate (p). A model that approximates sufficiently well TCP's steady-state throughput (T_h) is given by [298]:

$$T_h(RTT, RTO, SMSS, p) = min \left\{ \frac{W_{max}}{RTT}, \frac{SMSS}{RTT\sqrt{\frac{2bp}{3}}} + RTO, \right.$$

$$\left. min \left\{ 1, 3\sqrt{\frac{3bp}{8}} p(1 + 32p^2) \right\} \right\} \qquad (3.3)$$

where b is the number of packets acknowledged in each ACK and W_{max} is the maximum size of the congestion window.

Remark 3.1 *The definition of TCP-friendliness assumes that all TCP flows are not treated unfairly by the non TCP-flows sharing the same congested link. However, in reality, such a condition is less likely to hold if not enforced by the flow control.*

Remark 3.2 *The throughput model (3.3) assumes that RTT and loss rate are independent of the source transmission rate. Therefore, (3.3) operates satis-factorily in environments with a high level of statistical multiplexing such as the Internet. Care should be taken though when used as part of a protocol's control loop where only a few flows share a bottleneck link.*

3.5.2   Classification of Congestion Control Protocols

Congestion control protocols are classified into many categories according to a number of operational features.

3.5.2.1   Window-Based vs. Rate-Based

Depending on the methodology followed to vary transmission rate, a congestion control protocol is classified either as rate-based or as window-based.

In window-based protocols, each source may transmit a specific maximum number of packets (defined through cwnd) before any new feedback (the ACK issued by the destination) arrives. The size of cwnd increases in the absence of congestion and decreases when congestion occurs. The destination accepts and counts all incoming packets and informs the source if it is allowed to increase the window and by which amount. Since source behavior is very strictly dictated by the presence or absence of incoming feedback, window-based control is said to be self-clocking.

Rate-based protocols adapt their transmission rate according to some integrated feedback algorithm that notifies about congestion when existent. Rate-based algorithms can be subdivided into simple AIMD mechanisms and model-based congestion control. Typically AIMD schemes result in saw-tooth throughput shape, making this type of protocol not fully compatible with the constraints imposed by streaming media applications. To smooth out source transmission rate, model-based algorithms use a TCP model to control congestion instead of the TCP-like AIMD, making them more suitable for the media traffic type, and consequently more TCP-friendly over a long-term scale. However, model-based congestion control may not resemble the TCP mechanism.

Current efforts adjust source transmission rate to ensure a minimum level of fairness between TCP and non-TCP flows.

3.5.2.2 Unicast vs. Multicast

Depending on the number of destination nodes a single source is transmitting packets simultaneously, the congestion control protocols are classified either as unicast or as multicast.

Multicast refers to the delivery of information to a group of destination nodes simultaneously, from a single source. On the contrary, in unicast, a source communicates with a single destination node.

Both unicast and multicast protocols need to be TCP-friendly. However, owing to the heterogeneity caused by variations in the network conditions at the destination side, the development of multicast protocols is far more complex than unicast. The complexity becomes severe as the number of destination nodes increases.

Single-rate is typically the mechanism adopted by all unicast congestion control protocols. Transmission in unicast has only one recipient, so it adapts its sending rate according to this recipient's status. Multicast transmission may adopt the single-rate approach as well, where the source streams data with the same rate to all recipients of the multicast group. This rate is chosen to suit the bottleneck of one of the receivers, but meanwhile it may not be the right one for others suffering less congestion. Hence, this approach limits the scalability of the multicast protocol.

Multi-rate transmission increases scalability in a multicast communication by assigning to each of the multicast session routes a specific transmission rate according to current network conditions. Multi-rate congestion control uses the layered multicast approach, since multi-layering enables the source to divide data into different layers to be sent to different multicast groups. Every destination has to join the largest possible number of groups permitted by the bottleneck. The quality of data sent to this receiver is higher when joining more groups. This feature is most obvious in multicast video sessions, where the more groups the recipient subscribes to, the more layers the recipient destinations, and the better the quality of video is. Meanwhile, for other bulk data, additional layers decrease the transfer time. By using this mechanism, congestion control is achieved implicitly through the group management and routing mechanisms of the underlying multicast protocol. A survey on multicast congestion control protocols can be found in [101].

3.5.2.3   End-to-End vs. Router-Based

Most of the congestion control algorithms are designed to work on an end-to-end basis that needs no support from the network to realize TCP-friendliness. End-to-end congestion control schemes have the advantage of being easily implemented on today's Internet. This category can be subdivided into source-based and destination-based schemes.

In source-based schemes, each source tunes its window size according to the information it receives regarding network congestion, to realize TCP-friendliness. In this formulation, only the destination nodes can send feedback. The rate tuning decision is totally seen as a responsibility of the source. On the other hand, in the destination-based congestion control, it is the destination's decision to subscribe or unsubscribe from extra layers, according to the network congestion status.

Generally, end-to-end schemes treat the network as a black-box. Therefore, end-to-end congestion mechanisms mainly rely on implicit congestion signals, i.e., packet losses and delay variations. Even though this approach has proved successful for many years, it nevertheless has some limitations. Mechanisms based in packet losses cause network congestion by design, as it is the only way to probe for available bandwidth, making them reactive, rather than proactive. Furthermore, mechanisms based on packet delay respond proactively to network congestion, and are able to keep packet losses and queueing delay at low levels. However, they face the dominant problem of the accurate queueing delay estimation.

Recently, end-to-end mechanisms have been developed that incorporate both loss and delay measurements [187], as well as others that use more sophisticated feedback signals [275], [274], leading to improved, yet far from optimum, results.

The mission of congestion control, especially the part of fairly sharing the network resources, can be facilitated by implementing part of its mechanisms over the network devices and not only at its terminals. Schemes that rely on an embedded functionality in the connecting routers are called router-based (or link-based). Multicast protocols in particular, depend on some of the network information regarding round-trip times and management of groups of receivers. Modification of the routers' queuing mechanism can greatly help the design and implementation of a multicast congestion control scheme. End-to-end schemes have the disadvantage of being dependable on the systems implemented on Internet terminals, allowing users to acquire all the available bandwidth via applications using non-TCP-friendly mechanisms. Router-based mechanisms however drop the non-TCP-friendly packets with a probability higher than that of dropping the TCP-friendly packets.

In early attempts, router-based mechanisms drop packets before routers reach the state of congestion. This approach has shown good results in controlling packet drops and queueing delay [37]. However, TCP sources still detect network congestion by observing packet losses and inherit all the problems associated with implicit congestion signaling.

Further attempts led to the development of mechanisms that use an explicit signal (utilizing one or two bits in the packet header) to notify the TCP sources, allowing them to adapt to congestion before it occurs and avoid unnecessary retransmissions. Such a modification has been shown to improve TCP performance but fails to solve all the problems, mainly owing to the limitations of the explicit feedback signaling.

These limitations have motivated researchers to explore more sophisticated algorithms for Internet congestion control which enable explicit multibit feedback. These algorithms allow routers to present sources with how much data they should be sending at a given time, therefore improving performance related to efficient utilization in high-speed long-distance links, fairness and the ambiguity of packet loss in heterogeneous networks.

Router-based has the disadvantage of being costly to deploy over the widely spread infrastructure of today's Internet, in terms of money, time and effort.

3.6 Concluding Comments

This book focuses on unicast single rate congestion control mechanisms able to support flow control at sources which are entrusted with real-time applications. The main purpose of a congestion control mechanism is to achieve high network utilization, small amounts of queueing delay and some degree of fairness among flows. Furthermore, the introduction of new types of services in the Internet establishes new quality requirements (Quality of Service) at congestion control mechanisms that should be maintained even under congested network conditions.

Congestion affects negatively the quality and reliability of applications, making imperative the use of modern congestion control mechanisms capable of regulating the round trip time of each packet, while preventing either overflows or empty queues, simultaneously guaranteeing fairness among competing sources. Moreover, modern applications seek cooperation with congestion control modules to provide real-time services under any network condition.

The implementation of the router based approach in the current Internet is very challenging owing to both technical and economical aspects. Typically, router based schemes significantly increase the complexity of the algorithms to be executed, which additionally require universal acceptance and enforcement in all network routers. It is widely accepted that a necessary requirement characteristic for a candidate link algorithm to be implementable is its simplicity. In other words the total number of computations performed owing to congestion control in a router should be kept to a minimum, a property of great significance especially when high-speed networks are considered.

All aforementioned drawbacks of router based congestion control algorithms moved research towards end-to-end (source based) schemes, reinforced with simple explicit feedback signaling, issued by the routers in the path, to notify the sources about the network operating conditions, leading to improved performance regarding utilization, delay and fairness, even in heterogeneous long-distance high-speed networks.

Such a configuration satisfies the requirement that the complexity of a network should be at its edges and not at its core (routers) [262], allowing the scaling of the network, without residing to router overload (with respect to computational complexity) having to modify only the mechanism with which a source transmits information.

4

Measuring Network Congestion

CONTENTS

4.1 Overview

End-to-end congestion control in computer networks, including the current Internet, requires some form of feedback information from the congested links to the sources of data traffic, so that they can adjust their rates of sending data according to the bandwidth available. Such feedback information can be transmitted either explicitly or implicitly.

In the case of implicit feedback, the network transport layer protocol tries to maintain high throughput and low delay of data packets by estimating the congestion level of the network, based on changes in end to end delay and packet drops. The TCP of the current Internet employs such an implicit feedback through timeouts and duplicate acknowledgments for lost packets. Specifically, TCP assumes that packet drops in a network occur mainly owing to congestion. This assumption is no longer valid, as wireless and satellite links are prone to transmission errors. Hence, all transport protocols that consider packet loss as an important factor in estimating congestion yield poor performance.

Other techniques used to estimate network congestion, rely on the variation of the packet round trip time (RTT) [162], [280]. However, comprehensive surveys have shown [203] that although there is useful congestion information contained within RTT samples, the level of correlation between an increase in RTT and packet loss is not strong enough to allow a source to reliably improve throughput.

The inability to determine the congestion level of a network based on changes in throughput, in end-to-end delay, as well as on packet drops, creates the need for the existence of an explicit feedback mechanism, according

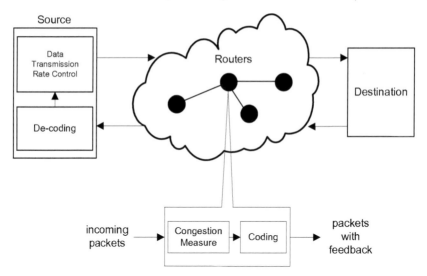

FIGURE 4.1
Feedback mechanism.

to which each router maintains a congestion price, which is a function of the buffer length and the arrival rate of incoming traffic and conveys the information about its calculated congestion price to the source.

In addition, keeping simple all algorithms executed on the links is imperative. Therefore, in what follows, we shall only present popular and low complexity proposals of explicit feedback mechanisms, analyzing their advantages and disadvantages. From a control system perspective, explicit feedback is equivalent to a measuring device. Its role is to measure path congestion, code this information in a format compatible to the underlying network technology and finally to transmit the information (through the network) back to the source of sending data. Figure 4.1 is illustrative.

4.2   Drop Tail

Drop tail [48] is the simplest congestion feedback algorithm and the most commonly used by routers in the current Internet. The name arises from the effect of the policy on incoming packets. Once a queue has been filled, the router begins discarding all additional packets, thus dropping the tail of the sequence of packets. The loss of packets causes the sources to reduce their sending rates until the routers recover and release their queues. Figure 4.2 illustrates the drop tail process on a router.

Simplicity, suitability to heterogeneity and decentralization are the main

incoming flows

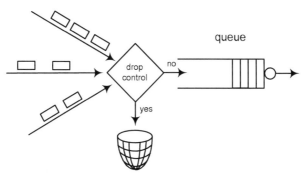

FIGURE 4.2
Drop tail process on a router.

advantages of drop tail. However, this approach has some serious drawbacks. More specifically, as there is no indication of congestion until packets from multiple TCP connections are dropped, the so-called global synchronization phenomenon [113], [99] is likely to appear, according to which all TCP sources involved enter slow start. This happens because, instead of discarding many packets from one connection, the router discards one packet from each connection.

Additionally, as data traffic is inherently bursty, routers are provisioned with fairly large buffers to absorb this burstiness and maintain high link utilization. In the case of drop tail these large buffers lead to high queueing delays at congested routers. Thus, drop tail forces network operators to choose between high utilization (requiring large buffers) or low delay (requiring small buffers).

In the problem of continuously transmitting high-bandwidth video and multimedia information, obtaining Quality of Service (QoS) guarantees on certain communication characteristics, including transmission rate and fairness, is closely related to the success of the task. In this direction drop tail proves inefficient. However, owing to its widespread deployment, drop tail is used as a baseline case for assessing the performance of any newly proposed router algorithm.

4.3 Congestion Early Warning

The reported weaknesses of drop tail are mainly owing to the lack of an early warning mechanism, capable of informing all sources involved, of the presence of an almost congested link in their paths. Such inside information could be used by the sources in the direction of reducing their data transmission rates, leading hopefully to congestion avoidance, or if unavoidable, to congestion escaping.

According to the means utilized to convey the early warning, the developed mechanisms are classified into two broad categories: a) packet drop, and b) packet marking schemes.

4.3.1   Packet Drop Schemes

Random Early Detection

The most significant representative of the class of packet drop schemes is random early detection (RED) [84], which uses queue size as a congestion measure. More specifically, in each router the average queue is constantly monitored and whenever congestion is detected, packets are dropped randomly.

For each packet arrival, the RED router calculates the average queue size (q_{avg}), using a low-pass filter with an exponential weighted moving average:

$$q_{avg}(k+1) = (1 - w_q)q_{avg}(k) + w_q q(k) \qquad (4.1)$$

where the parameter w_q determines the time constant of the low-pass filter and $q(k)$ is the queue length at the k instant. The calculated q_{avg} is compared with two thresholds, a minimum threshold (min_{th}) and a maximum threshold (max_{th}). When the average queue size is less than the min_{th}, no packets are dropped. When the q_{avg} is between the min_{th} and the max_{th}, each arriving packet is dropped with probability p_b, which is given by:

$$p_b(k) = p_{max} \frac{q_{avg}(k) - min_{th}}{max_{th} - min_{th}} \qquad (4.2)$$

where p_{max} is a RED parameter. As q_{avg} varies from min_{th} to max_{th}, the packet-marking probability p_b varies linearly from 0 to p_{max}. When the average queue size is greater than max_{th}, every arriving packet is dropped. Assuming that all sources are cooperative, the average queue size will not significantly exceed max_{th}.

To ensure that the router does not wait too long before marking a packet, it is proposed [84] to use p_a as the dropping probability, which increases slowly with the count variable, which indicates the number of packets forwarded since the last drop. Specifically

$$p_a(k) = \frac{p_b(k)}{1 - p_b(k)count}. \qquad (4.3)$$

A graph showing the dropping probability p_b versus the average queue length q_{avg} is presented in Figure 4.3.

The main disadvantage of RED is that its performance [184], [256], [277] is very sensitive to parameter settings. A badly configured RED will do no

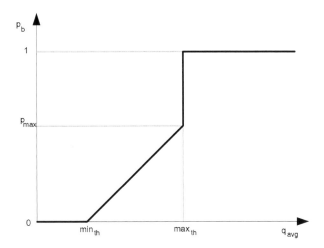

FIGURE 4.3
Dropping probability (p_b) versus the average queue length employed in RED.

better than drop tail. In that respect, the average queue parameter w_q is determined by the size and duration of bursts in queue size that are allowed at the gateway. The minimum and maximum thresholds min_{th} and max_{th} are determined by the desired average queue size.

Furthermore, it has been shown in [205] that RED operation results in large jitter in the buffers. Therefore, in CBR applications (e.g., audio streaming) any performance improvement on delay is almost completely compensated by the latency introduced.

In the original version of RED, all incoming packets are dropped if $q_{avg} > max_{th}$. This may lead to oscillatory behavior [76]. This problem is relaxed by avoiding the abrupt change of p_b to 1, as average queue size approaches max_{th}, replacing it by a gradual increase, as is pictured in Figure 4.4. Such a modification increases robustness with respect to RED parameter selection and to undesired oscillations in queue size.

Adaptive Random Early Detection (ARED)

As previously stated, tuning the RED parameters is certainly not a trivial task that strongly depends on network operating conditions [115] which, unfortunately, are unavailable to the designer. Therefore, it would be preferable, instead of having fixed values, to be able to carry out measurements and automatically update the RED parameters on the fly.

This thought brought to the foreground the adaptive RED (ARED) scheme. According to various versions presented, the primary objective is to adapt p_{max} in order to keep the average queue size between min_{th} and max_{th}. More specifically, the philosophy in altering p_{max} is summarized as follows [80]:

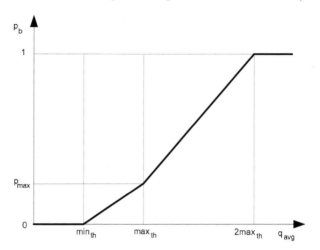

FIGURE 4.4
A modified dropping probability scheme that increases robustness of RED with respect to parameter selection and to undesired oscillations in queue size.

- the average queue size (q_{avg}) should be maintained within a target range half way between min_{th} and max_{th};

- p_{max} varies in a slower time scale than the typical round trip time;

- p_{max} should remain within $[0.01, 0.5]$;

- p_{max} should obey an additive-increase multiplicative decrease (AIMD) policy.

A specific implementation that fulfills the aforementioned guidelines is

$$p_{max}(new) = \begin{cases} p_{max}(old) + \alpha & \text{, if } q_{avg} > target \text{ and } p_{max} \leq 0.5 \\ p_{max}(old)\beta & \text{, if } q_{avg} < target \text{ and } p_{max} \geq 0.001 \\ p_{max}(old) & \text{, if } q_{avg} = target \end{cases}$$

where $target = (max_{th} + min_{th})/2$, $\alpha = min\{0.01, p_{max}/4\}$ and $\beta = 0.9$.

In RED parameters w_q controls the reactiveness of the average queue length to fluctuations of the instantaneous queue and strongly depends on link capacity. To control reactiveness ARED sets w_q to be a function of the link capacity. More precisely,

$$w_q = 1 - exp(\frac{1}{C}) \qquad (4.4)$$

where C is the link capacity. Finally, min_{th} and max_{th} are selected as

$$min_{th} = max\{5, target_{delay}\frac{C}{2}\} \tag{4.5}$$

$$max_{th} = 3min_{th} \tag{4.6}$$

where $target_{delay} = 5ms$.

The main advantage of ARED [164], [173] is that it works autonomously, setting its parameters in response to the changing traffic load, even though the selection may not be necessarily optimal.

BLUE

Conceptually different from RED, BLUE [75] assumes that queue length does not directly reflect the congestion level. Hence, it does not update the packet marking probability with the queue length.

Instead, it uses queue overflow and idle event history to update the packet marking probability (p_m). Packet loss owing to queue overflow means that the marking is not aggressive enough and p_m should be increased. Similarly, the queue idle event occurs as a result of the aggressive marking policy. Therefore, the p_m parameter should be decreased.

BLUE uses three parameters: The first two determine the amount by which p_m is incremented in case the queue overflows (d_i), or is decremented when the link is idle (d_d). The third parameter $(freeze\_time)$ is the minimum time interval between two successive updates. BLUE is a simple mechanism, compared to RED, and is mainly used to protect TCP flows against non-responsive flows.

CHOKe

In the CHOKe algorithm [223], whenever a new packet arrives at the congested router, a packet is drawn at random from the FIFO buffer, and is compared with the arriving packet. If both belong to the same flow, then both are dropped. Otherwise, the randomly chosen packet is kept intact and the new incoming packet is admitted into the buffer with a probability that depends on the level of congestion. This probability is computed as in RED.

It is truly a simple algorithm which does not require any special data structure, aiming at controlling unresponsive or misbehaving flows with a minimum overhead. However, CHOKe is not likely to perform well when the number of flows is large compared to the buffer space.

Adaptive Virtual Queue

The adaptive virtual queue (AVQ) scheme [171] differs from the other mechanisms that we have discussed thus far in that it does not explicitly calculate a marking probability. Instead, it maintains a virtual queue whose link capacity $\tilde{C}$ is less than the actual link capacity C ($\tilde{C} < C$) and whose buffer size is equal to the buffer size of the real queue. The algorithm tries to keep the link utilization to an area close to a desired link utilization γ. This is done by monitoring the arrival rate of packets and not the queue length (as RED does), providing in this way faster reaction [158]. Whenever a packet arrives, it is enqueued in the virtual queue if there is space available. Otherwise, the packet is dropped. The capacity of the virtual queue is updated at each packet arrival according to the differential equation

$$\dot{C} = \alpha(\gamma C - \lambda) \tag{4.7}$$

where λ is the arrival rate at the link and α is a damping factor that controls how fast the mechanism reacts. Both parameters determine the stability of the AVQ algorithm [189], [305].

4.3.2   Packet Marking Schemes

Packet dropping as an early warning congestion mechanism results in packet retransmission, which negatively affects the network traffic and increases packet delay.

Nevertheless, in applications where satisfying QoS metrics is crucial, packet dropping may be a prohibitive action. In marking bit schemes, packets are marked instead of dropped, alleviating in this way the aforementioned problems.

In TCP/IP networks, the Explicit Congestion Notification (ECN) protocol provides the mechanism for using the Congestion Experienced (CE) code point in the packet header as an indication of congestion. Specifically, ECN uses the two least significant (right-most) bits in the IP header to encode four different codepoints:

00: Non ECN-Capable Transport, Non-ECT
10: ECN Capable Transport, ECT(0)
01: ECN Capable Transport, ECT(1)
11: Congestion Encountered, CE

When both endpoints support ECN, they mark their packets with ECT(0) or ECT(1). If the packet traverses a router with a feedback mechanism (e.g., RED) that is experiencing congestion and the corresponding router supports ECN, it may change the codepoint to CE instead of dropping the packet. This act is referred to as "marking" and its purpose is to inform the receiving endpoint of impending congestion. At the receiving endpoint this congestion indication is handled by the upper layer protocol (transport layer protocol) and needs to be echoed back to the source in order to reduce its transmission rate.

TCP supports ECN using two flags in the TCP header. Those two bits are used to echo back the congestion indication and to acknowledge that the congestion-indication echoing was received. These are the ECN-Echo (ECE) and congestion window reduced (CWR) bits.

When ECN has been negotiated on a TCP connection, the sender indicates that IP packets that carry TCP segments of that connection are carrying traffic from an ECN capable transport by marking them with an ECT codepoint. This allows intermediate routers that support ECN to mark those IP packets

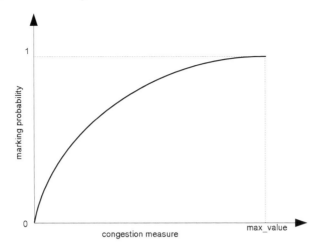

FIGURE 4.5
Marking probability in REM.

with the CE codepoint instead of dropping them in order to signal impending congestion.

Upon receiving an IP packet with the congestion experienced codepoint, the TCP destination echoes back this congestion indication using the ECE flag in the TCP header. When an endpoint receives a TCP segment with the ECE bit, it reduces its congestion window. It then acknowledges the congestion indication by sending a segment with the CWR bit set. A node keeps transmitting TCP segments with the ECE bit set until it receives a segment with the CWR bit set.

Apparently, ECN can only be used when supported from the source, the destination and the underlying network (all routers in the corresponding source-destination path).

Notice also that with ECN the congestion notification is promptly received by the TCP source. Therefore, the source-destination connection does not remain idle for a significant amount of time, waiting for the TCP retransmit timer to expire, which is common in the packet drop case.

In what follows specific marking policies will be presented.

Random Exponential Marking (REM)

According to REM philosophy [12], [308], at a local level (in every link l) a price variable p_l is maintained as a congestion measure, determining the packet marking probability. REM updates the price variable, based on the difference between input rate and link maximum bandwidth (rate mismatch) and the difference between queue length and a target value (queue mismatch). The update may be done periodically or asynchronously. Typically, p_l is increased, if the input rate is larger than the link maximum bandwidth or the queue size is larger than the corresponding target value, and decreased otherwise. A possible formula for p_l variation follows:

$$p_l(t+1) = max\{p_l(t) + \gamma(a_l(b_l(t) - b_l^*) + x_l(t) - c_l(t)), 0\} \qquad (4.8)$$

where γ and a_l are small positive constants, $b_l(t)$ is the aggregate buffer occupancy at queue l, $b_l^* \geq 0$ is the target queue length, $x_l(t)$ is the aggregate input rate to the queue and $c_l(t)$ is the available bandwidth. The difference $x_l(t) - c_l(t)$ measures the rate mismatch and the difference $b_l(t) - b_l^*$ measures the queue mismatch. The constant a_l trades off utilization and queueing delay during transient. The constant γ controls the responsiveness of REM to changes in network conditions. At the equilibrium point [92], [309], [317], the price stabilizes and $a_l(b_l - b_l^*) + x_l - c_l = 0$. This is true only if the mismatches are driven to zero ($x_l = c_l$ and $b_l = b_l^*$), yielding in this way high utilization, negligible packet losses and delay.

At a global level, REM uses the sum of all link prices along a path as a measure of path congestion, and determines an end to end marking probability which is what the source finally measures.

Each queue marks every packet (not already marked) in an upstream queue, with a probability that is exponentially increasing with the current price. This marking probability is illustrated in Figure 4.5.

Suppose a packet traverses links $l = 1, 2, ..., L$ that have prices $p_l(t)$. Then the marking probability $m_l(t)$ at queue l is

$$m_l(t) = 1 - \phi^{-p_l(t)} \qquad (4.9)$$

where $\phi > 1$ is a constant. The end-to-end packet marking probability is

$$M(t) = 1 - \prod_{l=1}^{L}(1 - m_l(t)) = 1 - \phi^{-\Sigma_l p_l(t)}. \qquad (4.10)$$

Therefore, the end-to-end marking probability $M(t)$, is high when the congestion measure of its path (i.e., $\Sigma_l p_l(t)$) is large [2].

When the link marking probabilities $m_l(t)$ are small, and hence the link prices $p_l(t)$ are small, the end-to-end marking probability given by (4.10) is approximately proportional to the sum of the link prices in the path [2]

$$M(t) = (log_e \phi) \sum_l p_l(t). \qquad (4.11)$$

The local computation requires no information aside from the link price, but the choice of ϕ is difficult. The performance of REM is highly dependent on the value of ϕ [2]. The optimal choice of ϕ requires knowledge of the path length and the end to end price, which generally cannot be deduced in a practical setting [269]. A suboptimal choice of ϕ may result in poor estimation performance.

Random Additive Marking (RAM)

RAM [2] marks packets with different probabilities at each router and uses side-information to calculate the sum of the link prices along a path. RAM achieves its goal provided two assumptions hold.

Assumption 4.1 *All link prices are bounded, so they can be normalized.*

This is a strong assumption, given the nature of congestion prices, which in principle, can be set to infinite values. However, this assumption is not as unrealistic as it might first appear. Prices may be explicitly bounded when they are defined in terms of a link cost function with bounded slope [98]. Even in cases where prices are not naturally bounded, it may be desirable to work with a truncated price range.

Assumption 4.2 *The source packet transmission protocol has knowledge of the total number of hops (intermediate routers) in the path.*

This feature can be implemented using the time-to-live (TTL) field in the IP header which is used to limit the maximum lifetime of a packet in the network. In addition to serving its intended purpose, the TTL field provides information regarding path lengths and thus could be plausibly used by a marking algorithm. Unlike the path length field that is initialized to zero and increased, TTL is initialized to some positive value and decreased. In [2] a TTL value estimator is proposed to transform the TTL field to a hop counter.

Suppose we restrict the range of each link price s_i to $0 \leq s_i \leq 1$, and suppose the step index i is known for local computation at the $i - th$ step. The initial marking state X_0, which is a bit in the ECN field, is set to 0.

At each step $i \geq 1$, link i leaves the price bit (X_i), unaltered $(X_i = X_{i-1})$ with a probability equal to $(i - 1)/i$, or it sets $X_i = 1$ with probability s_i/i and 0 otherwise.

The resulting X_n is a $0 - 1$ random variable with $E[X_n] = \Sigma_{i=1}^{n} s_i/n$. We thus have an unbiased estimator for z_n/n, where z_n is the total price along the path; we simply collect N samples and compute the average, $\overline{X}$. Since the step index is known at each step, the receiver can determine n and thus obtain z_n [2].

It is reported [2] that RAM outperforms even an optimally parameterized REM when the prices are uniformly distributed in terms of mean squared error. Furthermore, the difficulty in setting the parameter ϕ in REM makes it difficult to deploy in heterogeneous network environments [2]. In that respect RAM can be an alternative.

However, both RAM and REM may provide a biased sum of prices, as their estimates are based on probability and the estimated original TTL may not be accurate. Moreover, the performance of RAM [275], [274] suffers severely when the average link price strays substantially away from 0.5.

Deterministic Packet Marking

This scheme extends the "ECN marking" proposal to provide an efficient deterministic marking algorithm, which allows the routers on a path connecting the source and the destination to convey the sum of their quantized prices to the destination, using the value of the IPid.

In the algorithm of deterministic packet marking for congestion price estimation, a router quantizes its link price s_i ($0 \le s_i \le 1$) to n levels, yielding a b-bit binary number ($b = log_2 n$).

All data packets are mapped into the so-called probe types. Each probe type may be modified by two routers so that it carries the sum of two bits of equal significance of the quantized prices of the two routers. The key idea behind this scheme is that each data packet encodes the sum of a small subset of link price bits. A hash function of the IPid field determines the probe type of a packet. When a packet of probe type i is transmitted, the router marks the packet if bit i of the quantized price is 1. When the data packet arrives at the destination, it indicates the sum of the two bits. The destination can reconstruct the sum of the quantized prices by combining all partial sums.

To communicate the sum of prices along a path of at most h hops, the algorithm introduces h probe types for each bit position. Any probe type is defined by the pair (h_i, b_i), where h_i is a hop number and b_i is the number of a bit position. Following [2], each router determines its "number" from the TTL field of the IP header (in this case as TTL mod h). For probes of type (h_i, b_i) only router h_i will mark the packet, if bit b_i in its price is set. In that respect, the destination can determine the price of each hop in the path.

This algorithm fully uses two ECN bits. The combination 00 means that ECN is not supported. The other three combinations, 01, 10, and 11 are for different marking values. This allows the destination to obtain information from up to six routers along the path by a single ECN probe. This requires

$$T_{sum} = 2b\lceil h/6 \rceil \qquad (4.12)$$

probe types for a b-bit quantizer and a path including at most h hops. In other words, to reliably estimate the sum of prices, at least T_{sum} packets must be received.

Works have reported [275], [274] that this algorithm performs better than RAM at the same number of ECN probes. However, the quantization used by this algorithm may result in possibly large mean-squared error (MSE), regardless of the number of probes. Thus, RAM will eventually obtain a lower MSE if the number of probes is sufficiently large, assuming that the price stays constant.

4.4 Concluding Comments

The mechanisms described in this chapter are just a small subset of the active queue management (AQM) schemes that can be found in the literature. For instance, AVQ is not the only one that maintains a virtual queue [98]. Alternatives can be found in [24], [33], [54], [59], [120], [170], [222], [284], [287], [288], [304], [314] describing integrated approaches tackling simultaneously both the problems of queue management and source behavior. Also, just like TCP, AQM mechanisms have been analyzed quite thoroughly [123], [154], [198]. Other overviews that are somewhat complementary to the one found here are given in [16], [25], [116], [195], [250], [283].

5

Source-Based Congestion Control Mechanisms

CONTENTS

5.1 Overview

TCP is the basis of the Internet and has served its requirements remarkably well [301]. However, its inefficiency in modern networks stimulates continuous improvement to match its rapid evolution. TCP variants that arise from this process recognize different congestion signals and adopt different congestion window updating rules, aiming at achieving high throughput, stability and fairness in various scenarios.

The role of a source based congestion control mechanism is to avoid congestive collapse, which occurs if the requirements of the traffic in a network are larger compared to the network capacity and simultaneously to achieve high throughput, fair allocation of the network resources and maximum compatibility with TCP.

5.2   Traditional TCP

TCP Tahoe [139], TCP Reno [141], [140] and TCP NewReno [211] are the traditional versions of TCP congestion control protocols currently deployed in the Internet, and they have achieved great success in performing congestion avoidance and control. As already discussed in Chapter 3, traditional TCP is based on a congestion window mechanism, which may be visualized as a packet buffer for the TCP source, used to record the packets which so far have not been acknowledged by the destination. The relationship connecting source rate and the size of the congestion window is

$$R = \frac{\overline{W}}{\overline{RTT}} \tag{5.1}$$

where R is the average source rate, $\overline{W}$ is the average congestion window size and $\overline{RTT}$ is the average round trip time. From (5.1) it can be deduced that congestion at a bottleneck can be alleviated by dynamically altering the congestion window.

The key idea of traditional TCP is for a source to gently probe the network for spare capacity by linearly increasing its rate and exponentially reducing its rate when congestion is detected. Congestion detection is introduced through packet losses.

Typically, the congestion window size is initialized at 1 packet and the source increments its value by one every time it receives an ACK. This phase which is called slow start, doubles the window every round trip time. When the congestion window reaches a threshold, the source enters the congestion avoidance phase. At this phase, the congestion window increases by one packet in each round trip time (the increase is additive and is implemented in each ACK). The slow start threshold indicates the available network capacity and is updated upon receiving a packet loss notification. In every packet loss, the source sets the slow start threshold to half of the current congestion window size and reenters the slow start phase, retransmitting the lost packet.

The traditional TCP window update algorithm which responds to each ACK is given by

$$w(t+1) = \begin{cases} w(t) + \frac{1}{w(t)}, & \text{if no congestion} \\ \\ \frac{1}{2}w(t), & \text{otherwise} \end{cases} \tag{5.2}$$

where $w(t)$ is the current congestion window size and $w(t+1)$ is the updated congestion window size.

The aforementioned algorithm is implemented in the TCP Tahoe. The traditional TCP version is completed with the use of the fast retransmit/fast recovery phase, which is however implemented in the TCP NewReno version, aiming at recovering from a loss more efficiently. The congestion control algorithm in the current TCP is referred to as TCP Reno in this book. Further details regarding the operation of the traditional version of TCP congestion control protocols can be found in Chapter 3.

Although TCP Tahoe and TCP Reno worked remarkably well in the early days of internetworking, they proved inefficient to comply with the more strict operational requirements of the current networks which may combine wireless links, long-distance high-speed connections, nonhomogeneous routing methodologies, serving applications with quality of service demands. This has led researchers to develop a large number of alternative congestion control mechanisms, some of which are incremental TCP improvements, others rely on additional implicit feedback, while others require the explicit participation of routers.

In the following sections the most representative TCP congestion control schemes will be presented.

5.3 TCP Modifications for Networks with Large Bandwidth Delay Products

TCP Reno has performed remarkably well and is generally believed to have prevented severe congestion as the Internet scaled up by magnitudes in size and speed. It is also well-known, however, that as the bandwidth-delay product grows, TCP Reno eventually becomes a performance bottleneck. This is for the following reasons:

1) Linear increase by one packet per round trip time is too slow, and multiplicative decrease per loss event is too drastic.

2) Maintaining large average congestion windows requires an extremely small loss probability in equilibrium.

3) TCP Reno uses a binary congestion signal (packet loss), which introduces strong oscillations in the congestion window.

The aforementioned inherent limitations of TCP motivated researchers to propose new protocols especially designed to operate in long-distance high-speed network environments. In this subsection the most representative TCP congestion control schemes will be presented.

A successful high-speed transport protocol candidate should possess the following properties:

Scalability: High-speed protocols should scale up their throughput without requiring unrealistically low packet loss rates.

TCP friendliness: High-speed protocols should offer TCP-compatible performance at high loss rates.

Responsiveness: High-speed protocols should respond promptly to available bandwidth variations.

5.3.1   Scalable TCP (STCP)

Scalable TCP is a simple modification to the traditional TCP congestion control algorithm [160], which dramatically improves TCP performance [3], [214] in high-speed long-distance networks. Slow start remains unaltered however, in the congestion avoidance phase, scalable TCP updates the congestion window to each ACK received according to

$$w(t+1) = w(t) + a \tag{5.3}$$

where a is a constant satisfying $0 < a < 1$. Further, upon detecting congestion in a given round trip time, the congestion window is altered by

$$w(t+1) = w(t) - bw(t) \tag{5.4}$$

where $0 < b < 1$ is a design constant. The use of $a = 0.01$ and b = 0.125 was suggested [160] to be a good choice with respect to legacy traffic, bandwidth allocation, flow rate variance, convergence and stability.

The interesting property of scalable TCP is that the recovery time after a loss event is decoupled from the window size. This period in the case of traditional TCP depends on the window size and the round trip time, while in case of scalable TCP it depends only on the round trip time. This means that a scalable TCP source requires less time to saturate a link after a loss event, and this is achieved by increasing the rate exponentially, rather than linearly. This property allows scalable TCP to outperform traditional TCP protocols in high-speed long-distance networks [15], [132].

5.3.2   HighSpeed TCP (HSTCP)

HighSpeed TCP [84], [259] is a TCP modification that applies only when the congestion window is large. The underlying idea of HSTCP is to alter its congestion window to achieve a large window size in environments with realistic packet loss ratios. To illustrate the necessity of such an achievement, notice that the traditional TCP average congestion window in steady state w_{std} is given by:

$$w_{std} = \frac{\sqrt{1.5}}{\sqrt{p}} \tag{5.5}$$

where p is the packet loss ratio [115]. Assuming that the packet size is 1500 bytes and the round trip time is 100 ms, for a traditional TCP flow to maintain a sending rate of 7.2 Gbps it is required

$$\frac{1500 \times 8}{0.1} \frac{\sqrt{1.5}}{\sqrt{p}} = 7.2 \times 10^9$$

which results in packet loss ratio $p = 4.17 \times 10^{-10}$. This packet loss ratio is beyond the current limits of achievable fiber error rates.

Upon receiving an ACK while in congestion avoidance mode, the HSTCP window is increased according to:

$$w(t+1) = w(t) + \frac{a(w)}{w} \qquad (5.6)$$

and when a loss is detected through triple duplicate acknowledgments, the window is decreased by

$$w(t+1) = w(t) - b(w)w(t). \qquad (5.7)$$

In (5.6), (5.7) $a(w)$ and $b(w)$ are appropriate functions of w. Typically, large values of w correspond to large $a(w)$ and low $b(w)$ levels.

A specific implementation [115] of $a(w)$, $b(w)$ distinguishes two cases:
Case 1 ($w \leq w_{low}$): When the congestion window is smaller than or equal to a constant (w_{low}), the functions $a(w)$, $b(w)$ are defined as

$$
\begin{aligned}
a(w) &= 1 \\
b(w) &= \frac{1}{2}.
\end{aligned}
$$

Case 2 ($w > w_{low}$): When the current congestion window is greater than w_{low} then:

$$a(w) = \frac{2w^2 b(w)p(w)}{2 - b(w)} \qquad (5.8)$$

$$b(w) = \frac{log(w) - log(w_{low})}{log(w_{high}) - log(w_{low})}(d_{high} - 0.5)0.5 \qquad (5.9)$$

$$p(w) = \exp\left[\frac{log(w) - log(w_{low})}{log(w_{high}) - log(w_{low})}P + log(p_{low})\right] \qquad (5.10)$$

where $P = log(p_{high}) - log(p_{low})$.

In (5.8)-(5.10), w_{low}, w_{high}, p_{low}, p_{high} and d_{high} are positive constants. Their default values are 38, 83000, 10^{-3}, 10^{-7} and 0.1 respectively. For small congestion windows ($w \leq w_{low}$), HighSpeed TCP behaves exactly like the traditional TCP.

Similar to scalable TCP, the congestion window is updated dynamically as determined by (5.6)-(5.10). However, in HighSpeed TCP the rate of change depends nonlinearly on the current congestion window [202]; a property which for high windows constitutes a more aggressive mechanism.

HighSpeed TCP is reported in [84], [166], [227] to be friendly to traditional TCP flows, utilizing the available bandwidth in high-speed long-distance networks, recovering fast in the presence of packet losses. It has minimum implementation requirements (only the congestion control mechanism of the traditional TCP needs modification) with straightforward collaboration with other TCP options. However, it is also reported [213] to present slow response to available bandwidth variations.

5.3.3 BIC

HSTCP and STCP schemes suffer from severe RTT unfairness [302]. RTT unfairness for high-speed long-distance networks occurs distinctly with drop tail routers for flows with large congestion windows where packet loss can be highly synchronized. Binary increase congestion control protocol (BIC) adapts its window depending on its current size. BIC consists of two parts: the binary search increase and the additive increase.

Binary search increase. In this part, congestion control is treated as a searching problem in which the system generates a binary feedback, whenever at packet loss the current window is larger than the network capacity. The starting points for this search are the current minimum window size W_{min} and the maximum window size W_{max}. Usually, W_{max} is the window size just before the last fast recovery (i.e., where the last packet loss occurred), while W_{min} is the window size just after the fast recovery. The algorithm repeatedly computes the midpoint between W_{max} and W_{min}, sets the current window size to the midpoint and checks for feedback, in the form of packet losses. Based on this feedback, the midpoint is taken as the new W_{max} if there is a packet loss, and as the new W_{min} if not. The process repeats, until the difference between W_{max} and W_{min} drops below a preset threshold, called the minimum increment (S_{min}).

The binary search increase technique, allows bandwidth probing to be initially more aggressive, when the difference of the current window size from the target window size is large, and becomes less aggressive as this error reduces. A unique feature of the BIC protocol is that its increase function is logarithmic. As a consequence it reduces its increase rate as the window size approaches the saturation point. Typically, the number of lost packets is proportional to the size of the last increment before the loss. Thus BIC may lead to reduced packet losses. The main benefit of binary search is that it gives a concave response function, which meshes well with the additive increase strategy whose description follows.

Additive increase. Additive increase strategy, in conjunction with binary search increase, may result in faster convergence and RTT fairness. When the distance of the midpoint from the current minimum is excessively large, increasing the window size directly to that midpoint may add too much stress to the network. When the distance of the current window size from the target in binary search increase is larger than a prescribed maximum step, called the maximum increment (S_{max}), the window size is increased by S_{max} until the distance becomes less than S_{max}, at which time the window increases directly to the target. Thus, after a large window reduction, the strategy initially increases the window linearly, and then increases logarithmically.

Combined with a multiplicative decrease strategy, binary increase becomes close to pure additive increase under large windows. This is because a larger window results in a larger reduction in multiplicative decrease, and therefore, a longer additive increase period. When the window size is small, it becomes close to pure binary search increase, resulting in a shorter additive increase period.

Slow start. When the current window size grows past the current maximum window W_{max}, the binary search algorithm switches to probing the new maximum window, which is unknown (i.e., the window size where loss can occur is unknown).

BIC converges sufficiently fast because the time it takes to find the ideal window is logarithmic. On the other hand, since BIC is designed for high-speed long-distance networks, its rate jumps can be quite drastic, which may lead to instability. To alleviate this problem, if the new midpoint is too far away from the current window, BIC additively increases its window, in fixed size steps, until the distance between the midpoint and the current window becomes smaller than one such step. Additionally, if the window grows beyond the current maximum in this manner, the maximum is unknown, and BIC therefore seeks out the new maximum more aggressively with a slow-start procedure; this is called "max probing." Further details regarding BIC can be found in [302].

5.3.4 CUBIC

CUBIC [239] is an enhanced version of BIC in the direction of simplifying the BIC window control and improving its TCP friendliness and RTT fairness. Although BIC behaves sufficiently well in those metrics in current high-speed environments, its growth function can still be too aggressive for TCP, especially under short RTT or low-speed networks. Furthermore, the required different window control parts add complexity in protocol analysis and design. It is therefore necessary to develop simple window growth functions to enhance TCP friendliness while retaining the strong points of BIC and especially its stability and scalability.

The window growth function of CUBIC is a cubic function, whose shape is very similar to the growth function of BIC. More specifically, the congestion window of CUBIC (W_{cubic}) is determined by the following function:

$$W_{cubic} = C(t - K)^3 + W_{max} \tag{5.11}$$

where C is a scaling factor, t is the elapsed time from the last window reduction, W_{max} is the window size just before the last window reduction, and

$$K = \sqrt[3]{W_{max}\frac{\beta}{C}}$$

where β is a constant multiplication decrease factor applied for window reduction at the time of loss event (i.e., the window reduces to βW_{max} at the time of the last reduction).

The window grows very fast upon a window reduction, but as it gets closer to W_{max}, its rate of increase reduces. Around W_{max}, the window increment becomes almost zero. Above that, CUBIC starts probing for more bandwidth in which the window grows slowly initially, accelerating its growth as it moves away from W_{max}. This slow growth around W_{max} enhances the stability of the protocol, and increases the utilization of the network, while the fast growth away from W_{max} achieves the required scalability properties.

The cubic function improves the intra-protocol fairness among competing flows of the same protocol. Qualitatively speaking, two flows competing on the same end to end path, eventually converge to a fair share, as they both have the same multiplicative factor β. Therefore, a flow that corresponds to a larger W_{max} will reduce more, while its increase rate will be slower, since K increases with W_{max}.

The CUBIC function also offers a good RTT fairness property, because the window growth rate is dominated by the elapsed time (t). This leads to linear RTT fairness, since any competing flows with different RTT will have the same t after a synchronized packet loss (note that TCP and BIC offer square RTT fairness in terms of throughput ratio). To further enhance fairness and stability, the window increment is clamped to S_{max} packets per second. This feature keeps the window growing linearly when it is far away from W_{max}, making the growth function very much in line with BIC's.

5.4   Delay-Based Congestion Control

When congestion occurs in a network, the queue of the bottleneck link increases and eventually, if its size is not appropriately compensated, the queue overflows and packets are dropped. The aforementioned phenomenon is directly related to the constant enlargement of the monitored queueing delay. It therefore seems reasonable to react upon an increasing delay. Note that only changes can be used as two mildly congested queues may yield the same delay as a single congested queue. Queueing delay is indirectly incorporated to TCP control via estimating the RTT.

Using delay measurements for congestion control, has the potential advantage of making more reasonable decisions, as the feedback from the network is enriched with further information (compared to the packet loss case). At the same time we should not underestimate the fact that in certain cases congestion is very loosely correlated to delay (e.g., in wireless networks). Nonetheless, two mildly congested queues may appear equivalent, in terms of delay, with a single congested queue.

In this section two very popular delay based congestion control mechanisms are presented. The first, TCP Vegas, was invented to compete with the traditional TCP Reno, while the second, FAST TCP, alleviates the weaknesses that occur in high-speed long-distance networks.

5.4.1 TCP Vegas

TCP Vegas [39], [197], [300] utilizes the packet queueing delay as a congestion signal to determine the rate at which to send packets. The estimate of queueing delay is calculated as the current RTT ($RTT(t)$) minus the actual round trip propagation delay. The latter results from the minimum RTT (RTT_{min}). The idea is to have a source estimate the number of its own packets buffered in the path and try to keep this number between α and β by adjusting its window size [251]. Where the positive parameters α and β are used as thresholds [209]. The window size is increased or decreased linearly in the next round trip time according to whether the current estimate is less than α or greater than β. Otherwise, the window size is unchanged.

The congestion window updating rule of TCP Vegas is given by

$$w(t+1) = \begin{cases} w(t) + \frac{1}{RTT(t)}, & \text{if } \frac{w(t)}{RTT_{min}} - \frac{w(t)}{RTT(t)} < \alpha \\ w(t) - \frac{1}{RTT(t)}, & \text{if } \frac{w(t)}{RTT_{min}} - \frac{w(t)}{RTT(t)} > \beta \\ w(t), & \text{otherwise} \end{cases} \quad (5.12)$$

where $w(t)$ and $w(t+1)$ are the current congestion window size and the updated congestion window size, respectively. The update of congestion window is performed at every RTT. Moreover, like TCP Reno, upon loss the congestion window is halved.

The significance of TCP Vegas is that it was the first that used RTT, as well as packet loss, as congestion signals. This enables TCP Vegas to calibrate its congestion window size more smoothly, so that bursts owing to abrupt increases of congestion window size and low utilization caused by window halving, to be substantially prevented. The congestion window remains unaltered when the difference between the expected and the actual rate is between the two thresholds. This is a major difference between TCP Vegas and TCP Reno, as the latter always needs to exceed the available capacity in order to detect congestion. TCP Vegas can detect incipient congestion and react early.

Regardless of its advantages, TCP Vegas has several unsatisfactory features which prevent its large-scale deployment [40], [88], [114]. TCP Vegas is a "polite" protocol which concedes to its contenders, such as TCP Reno. Another shortcoming is the inaccurate measurement of the actual round trip propagation delay. The algorithm depends heavily on accurate calculation of the actual round-trip propagation delay. If it is too small, then the throughput of the connection will be less than the available bandwidth, while if the value is too large, then it will overrun the connection. A lot of ongoing research is currently targeted at the fairness of the linear increase/decrease mechanism of the congestion control of TCP Vegas [36], [44], [87], [268]. One interesting caveat is when TCP Vegas is inter-operated with other TCP versions like Reno [272], [279]. In this case, the performance of TCP Vegas degrades because Vegas reduces its sending rate before Reno, as it detects congestion early and hence gives greater bandwidth to co-existing TCP Reno flows.

5.4.2   FAST TCP

In high-speed long-distance networks, mechanisms like HighSpeed TCP and STCP are limited by the oscillatory behavior of TCP [93], [143] [150], [151], [310] which is an unavoidable outcome of binary feedback from packet loss. The problem remains even in the presence of ECN because a marked packet is interpreted just like a single packet loss event. In such networks, queueing delay can be much more accurately sampled because loss events are very rare, and loss feedback, in general, has a coarse granularity.

FAST has been called the high-speed version of Vegas, as it also takes the relationship between the minimum RTT and the recently measured RTT into account.

FAST operation is comprised by four modules [150]:

- Data Control

- Window Control

- Burstiness Control

- Estimation.

The Data Control determines which packets to transmit, Window Control determines how many packets to transmit, and Burstiness Control determines when to transmit these packets. The Estimation module drives the other parts by providing necessary information. Window Control regulates packet transmission at the RTT timescale, while Burstiness Control works at a smaller timescale. Below is the description of the Estimation and Window Control modules.

Estimation. This unit provides estimations of various parameters which are inputs to the other three decision-making modules. It computes two pieces of feedback information for each data packet sent. From every ACK that arrives, it takes a RTT sample and updates the average queueing delay and the RTT_{min}. When a negative acknowledgment (signaled by three duplicate acknowledgments or a timeout) is received, it generates a loss indication for this data packet to the other modules. The estimation unit generates both a multi-bit queueing delay sample and a single-bit loss-or-no-loss sample for each data packet.

The queueing delay is smoothed by taking a moving average with the weight $\eta(t)$ that depends on the window $w(t)$ at time t, as follows

$$\eta(t) = min\left\{\frac{3}{w(t)}, \frac{1}{4}\right\}. \tag{5.13}$$

The $k-th$ RTT sample $RTT(k)$ updates the average RTT, $(\overline{RTT(k)})$ according to:

$$\overline{RTT(k+1)} = (1 - \eta(t_k))\overline{RTT(k)} + \eta(t_k)RTT(k) \qquad (5.14)$$

where t_k is the time at which the $k-th$ RTT sample is received. Taking $RTT_{min}(k)$ to be the minimum RTT observed thus far, the average queueing delay is estimated as:

$$\hat{q}(k) = \overline{T(k)} - RTT_{min}(k). \qquad (5.15)$$

Window control. The window based FAST TCP is updated reacting to both queueing delay and packet loss. Under normal network conditions, FAST periodically updates the congestion window based on the average RTT $(\overline{RTT})$ and the average queueing delay $(\hat{q})$ provided by the estimation unit, according to:

$$w(t+1) = \min\left\{ 2w(t), (1-\gamma)w(t) + \gamma\left(\frac{RTT_{min}}{RTT}w(t) + a(w(t), \hat{q})\right)\right\} \qquad (5.16)$$

where $\gamma \in (0,1]$ and a is a function of the current window size and the average queuing delay. The constant a produces linear convergence when the queuing delay is zero [150], [167], [270].

FAST TCP is shown [57], [58], [91], [286], [285], [312] to be stable and to converge to weighted proportional fairness, assuming that all users have a logarithmic utility function. Additionally, it is reported [293] to perform remarkably well in numerous simulations and real-life experiments, thereby making a good case for the inclusion of additional fine-grain feedback in the TCP response function.

5.5 Congestion Control for Wireless Networks

The traditional TCP implementations rely on packet loss as an indicator of network congestion and lack the ability to distinguish congestion losses from losses invoked by noisy links. In wireless connections, overlapping radio channels, signal attenuation and additional noises have a huge impact on such losses. As a consequence, the traditional TCP reacts with drastic reduction of the congestion window, thus degrading its performance.

This section gives an overview of some modifications that have been proposed to improve TCP performance over wireless networks.

As TCP interprets all losses as congestion in the network, various explicit notification schemes have been proposed to enable the TCP source to identify the real cause of a data loss. The idea is that an explicit notification is transmitted in order to inform the source about data loss owing to corruption in the wireless network.

End to end proposals make the TCP source handle packet losses caused by both congestion and random wireless errors and requires minimal or no processing at the base station. Another advantage of these schemes is that the end-to-end semantics of TCP is maintained.

5.5.1   TCP Westwood

TCP Westwood [97] is based on end to end bandwidth estimation to set the congestion window and slow start threshold after a congestion episode, i.e., after three duplicate acknowledgments or a timeout. TCP Westwood, instead of applying the fixed rate decrease by half, it sets this variable equal to the product of the estimated bandwidth and the minimum RTT. This leads to a more-drastic reduction in the case of severe congestion than in the case of light congestion.

Each source performs an end-to-end estimate of the available bandwidth along a TCP connection by measuring and averaging the rate of returning ACKs. After a congestion episode, the source uses the measured bandwidth to properly set the congestion window and the slow start threshold.

In that respect, if an ACK is received at the source at time t_k, this implies that a corresponding amount of data d_k has been received by the TCP destination. Therefore, the following sample of bandwidth used by that connection is calculated:

$$b_k = \frac{d_k}{t_k - t_{k-1}} \tag{5.17}$$

where t_{k-1} is the time the previous ACK was received. Since congestion occurs whenever the low-frequency input traffic rate exceeds the link capacity, we employ a low-pass filter to average sampled measurements and to obtain the low-frequency components of the available bandwidth. Notice that this averaging is also critical to filter out the noise owing to delayed acknowledgments.

The choice of the filter is important. A simple exponential averaging of the kind used by TCP for RTT estimation is unable to efficiently filter out high-frequency components of the bandwidth measurements. The following discrete time filter has been proposed [97], which is obtained by discretizing a continuous low-pass filter using the Tustin approximation:

$$\hat{b}_k = \frac{\frac{2\tau}{t_k - t_{k-1}} - 1}{\frac{2\tau}{t_k - t_{k-1}} + 1} \hat{b}_{k-1} + \frac{b_k + b_{k-1}}{\frac{2\tau}{t_k - t_{k-1}} + 1}. \tag{5.18}$$

In 5.18 $\hat{b}_k$ is the filtered measurement of the available bandwidth at time $t = t_k$, and $1/\tau$ is the cut-off frequency of the filter. When TCP Westwood determines that the link is congested after receiving three duplicate ACKs, it sets the ssthresh to its estimated bandwidth delay product as follows:

$$ssthresh = \frac{\hat{BW}\ RTT_{min}}{p_{size}} \tag{5.19}$$

where $\hat{BW}$ is the estimated bandwidth, RTT_{min} is the minimum of TCPs estimation of round trip delay, and p_{size} is the TCP packet size in bits. The congestion window, cwnd, is then set equal to ssthresh if the connection is in the congestion avoidance mode.

TCP Westwood has been shown [47], [97], [96], [51] to give improvements over wireless as well as over wired links. Error recovery is faster than in traditional TCP, since the bandwidth estimation is considered when values for cwnd and ssthresh are computed after data loss.

5.5.2 TCP Veno

TCP Veno [89] is an end-to-end congestion control mechanism that is simple and effective for dealing with random packet losses. A key ingredient of TCP Veno is that it monitors the network congestion level and uses that information to decide whether packet losses are likely to be owing to congestion or to random bit errors.

TCP Veno combines ideas from Vegas and Reno using a mechanism similar to that in TCP Vegas to estimate the connection state, retaining the essential idea of Reno, in which the window size is increased progressively in the absence of packet loss. If a packet loss is detected while the connection is in the congestive state, TCP Veno assumes the loss is owing to congestion; otherwise, it assumes the loss is random.

The source measures the so-called expected (r_e) and actual (r_a) rates as follows:

$$r_e = \frac{w}{RTT_{min}} \tag{5.20}$$

$$r_a = \frac{w}{RTT} \tag{5.21}$$

where RTT is the smoothed round-trip time measured.

When $RTT > RTT_{min}$, there is a bottleneck link where the packets of the connection accumulate. Let the backlog at the queue be denoted by N. Then,

$$RTT = RTT_{min} + \frac{N}{r_a}. \tag{5.22}$$

Rearranging, we obtain

$$N = r_a(RTT - RTT_{min}) = e_r RTT_{min} \tag{5.23}$$

where $e_r = r_e - r_a$.

TCP Veno algorithm uses N (backlog packets in the path buffers) as an indication of whether the connection is in congestive state. If N is greater than a constant β[1] when a packet loss is detected, TCP Veno assumes that

---

[1] Experiments [89] indicate that $\beta = 3$ is a good setting.

the loss is random and not owing to congestion, and a different window adjustment scheme from that in Reno will be used; otherwise, TCP Veno assumes a congestion originating loss, and the Reno window adjustment scheme will be adopted. More specifically, TCP Veno modifies the fast retransmit, (called fast recovery) of Reno as follows.

In the presence of packet loss, if $N < \beta$ TCP Veno reduces the *ssthresh* according to

$$ssthresh = \xi w$$

where $\frac{1}{2} < \xi < 1$ so that the cutback in window size is less drastic than the case when loss is owing to congestion. In turn, the window size reduces to

$$w = ssthresh + 3.$$

If $N \geq \beta$, TCP Veno adjusts its window according to

$$w_{t+1} = w_t + \frac{1}{w_t} \tag{5.24}$$

which is the additive increase formula of TCP Reno, with the exception that it is operated on a different time scale. In fact TCP Veno increases its congestion window by one packet every two round trip times.

Experiments show [89], [311] that TCP Veno presents fewer oscillations than Reno does. In typical wireless access networks with 1% random packet loss rate, a throughput improvement of up to 80% is reported [224], [89], [313], [316], [315]. A salient feature of TCP Veno is that it modifies only the source side protocol of Reno, without changing the destination side protocol stack.

5.6   Congestion Control for Multimedia Applications

Nowadays, there is a significant increase in internet traffic from multimedia applications. Compared to classical applications, real-time media applications have unique characteristics such as stable round trip time per packet. Congestion control mechanisms should present smooth congestion response in order to achieve round trip regulation.

In this section are presented two very popular congestion control mechanisms for multimedia applications. The first, RAP, is an end to end rate-based congestion control mechanism, and the second, TFRC, is an equation-based congestion control scheme.

5.6.1   Rate Adaptation Protocol (RAP)

The rate adaptation protocol (RAP), uses a congestion control mechanism which mimics the AIMD behavior of TCP in a rate-based fashion [238]. A

RAP source sends data packets with sequence numbers, and a RAP destination acknowledges each packet, providing an acknowledgment (ACK) as feedback. Each (ACK) packet contains the sequence number of the corresponding delivered data packet.

On the basis of this feedback, the source can detect packet losses and sample the round trip time (RTT). Packet loss is detected via three DUPACKs (reception of three packets after one that is missing). In addition, there is a timeout mechanism as in TCP. Finally, the source explicitly calculates and controls the time between sending two packets, which is called the inter-packet-gap (IPG).

To design a rate adaptation mechanism, three issues must be addressed:

- the decision function;

- the increase/decrease algorithm;

- the decision frequency.

Decision Function. The rate adaptation philosophy can be summarized as follows: a) if no congestion is detected, periodically increase the transmission rate and b) if congestion is detected, immediately decrease the transmission rate.

RAP considers losses to be congestion signals, and uses timeouts and out of order packets to detect a loss. RAP maintains an estimate of RTT, denoted by $\hat{RTT}$ and calculates the timeout based on the Jacobson/Karel's algorithm [142].

Increase/decrease Algorithm. RAP uses an additive increase - multiplicative decrease (AIMD) algorithm to adapt its rate. In the absence of packet loss, the transmission rate r is periodically increased in a step-like fashion, using a "step height" α. More specifically, at each adjusting point the transmission rate is given by

$$r_{i+1} = r_i + \alpha \tag{5.25}$$

where $\alpha = \frac{p_{size}}{C}$, with C a positive constant. Given the IPG, the transmission rate is

$$r_i = \frac{p_{size}}{IPG_i}. \tag{5.26}$$

Therefore it is directly controlled by adjusting the IPG. Substituting (5.26) into (5.25) it is obtained

$$
\begin{aligned}
r_{i+1} &= \frac{p_{size}}{IPG_i} + \frac{p_{size}}{C} \\
&= p_{size} \left[\frac{C + IPG_i}{C\, IPG_i} \right].
\end{aligned} \tag{5.27}
$$

An iterative formula for updating the IPG is provided by combining (5.26) and (5.27). Hence,

$$IPG_{i+1} = C\frac{IPG_i}{C + IPG_i}. \tag{5.28}$$

Upon detecting congestion, the transmission rate is decreased multiplicatively. This is achieved by doubling the value of IPG. Hence,

$$IPG_{i+1} = 2IPG_i \tag{5.29}$$

and (5.26) yields

$$r_{i+1} = \frac{p_{size}}{IPG_{i+1}} = \frac{1}{2}r_i. \tag{5.30}$$

Decision Frequency. Decision frequency specifies how often to change the transmission rate. The optimal adjustment frequency depends on the feedback delay. Feedback delay in ACK-based schemes is equal to one RTT. It is suggested that rate-based schemes adjust their rates not more than once per RTT [201]. Changing the rate too often results in oscillations, whereas infrequent changes leads to unresponsive behavior.

RAP adjusts the IPG once every $\hat{RTT}$ using (5.28). The time between two subsequent adjustment points is called a step. If no loss is detected, IPG is decreased and a new step is started. Adjusting the IPG once every $\hat{RTT}$ has a nice property; packets sent during one step are likely to be acknowledged during the next step. This allows the source to observe the reaction of the network to the previous adjustment before making a new decision. If the value of IPG is updated once every $\hat{RTT}$ and the value of C is chosen to be equal to $\hat{RTT}$, the number of packets sent during each step is increased by 1. Since the length of each step is $\hat{RTT}$ and the height of each step is inversely proportional to $\hat{RTT}$, the slope of the transmission rate is also inversely proportional to $\hat{RTT}$.

RAP is a simple, TCP-friendly, rate-based control scheme. It has been reported [238] that it can be relatively easily embedded in an end-to-end architecture for real-time adaptive multimedia data transmission with intelligent buffer control, with smoothness being not a critical issue, however.

TCP's slope of linear increase is related to RTT in the same way as in the steady state. Thus, a RAP source can exploit RTT variations and adaptively adjust its rate in the same manner as TCP. In that respect, the adaptive rate adjustment in RAP emulates the coarse-grain rate adjustment in TCP.

It has been reported [78] that RAP connections with shorter RTTs are more aggressive and achieve a larger share of the bottleneck bandwidth. Therefore, RAP appears unfair to flows with larger RTTs.

5.6.2   TFRC

TCP friendly rate control (TFRC) is a congestion control scheme whose primary goal is to maintain a steady-like sending rate, while still being responsive to congestion. It was originally designed for multimedia streaming, but it can be also used by applications requiring smooth rate transmission (i.e., VoIP).

To compete fairly with TCP, TFRC uses the TCP throughput equation, which roughly describes TCP's sending rate as a function of the loss event rate, the round trip time, and the packet size. A loss event is defined as one or more lost or marked packets from a window of data, where a marked packet refers to the congestion indication of the explicit congestion notification (ECN) mechanism. The loss event rate calculation at the destination is perhaps the most critical part of TFRC.

TFRC sources transmit packets at a rate calculated on the basis of a detailed model for the rate of a TCP connection. This rate is determined by the congestion that exists on the network through the packet feedback mechanism that is periodically issued by the destination.

Details regarding the TFRC congestion control scheme can be found in [52], [81], [152], [177], [182], [257], [299], [303]. The operation of TFRC is characterized by its smooth sending rate, which comes at the cost of responsiveness.

5.7 Concluding Comments

This chapter is devoted to the presentation of representative source-based congestion control mechanisms including TCP and its variants which were reported to extend the applicability of the traditional TCP (currently deployed) to network environments that may include long-distance high-speed and/or wireless connections, that may intend to support quality demanding applications. Table 5.1 summarizes their basic characteristics and benefits.

Emphasis was placed on qualitative operational aspects, leaving the study of more control theoretic issues, like stability, convergence to equilibriums, robustness to uncertainties and network variations, to subsequent chapters. It is important to underline however, that for the majority of the presented congestion control protocols the aforementioned issues still constitute open problems.

TABLE 5.1

Qualitative presentation of source-based congestion control mechanisms

Mechanism	Feedback	Benefit
New Reno	Packet Loss	-
Vegas	Delay	Reduced packet losses
HighSpeed	Packet Loss	Applies to high-speed long-distance networks
BIC	Packet Loss	Applies to high-speed long-distance networks
CUBIC	Packet Loss	Applies to high-speed long-distance networks
STCP	Packet Loss	Applies to high-speed long-distance networks
FAST	Delay	Applies to high-speed long-distance networks
TCP Veno	Packet Loss/Delay	Applies to networks with wireless connections
Westwood	Packet Loss/Delay	Applies to networks with wireless connections
RAP	Packet Loss	Smooth rate achievement
TFRC	Packet Loss	Smooth rate achievement

6

Fluid Flow Model Congestion Control

CONTENTS

6.1 Overview

Current Internet is comprised of a plethora of heterogeneous systems (i.e., sources, routers) performing highly complex tasks for which it is practically impossible to obtain clear and accurate mathematical representations. In addition, the interconnected medium (wired, wireless, satellite links) elevates even further the complexity level of the overall system. To confront the extreme heterogeneity - complexity, Internet congestion control has relied on engineering solutions, mostly of heuristic nature, some of which were presented in Chapter 5. The lack of a coherent explanation of why all these methods work and, under which conditions, fail, defines their most significant deficiency. To deal with this issue, further complicated solutions are proposed, leading to a spiral of increasing complexity.

In this direction, simulations and experiments in real networks have revealed [1], [31], [53], [104], [172], [183], [219], [253] that as the bandwidth - delay product tend to increase, the additive increase - multiplicative decrease (AIMD) policy of the TCP protocol leads to system underutilization. Furthermore, analytical studies have shown [5], [17], [21], [109], [196], [233], [242] that in such cases TCP becomes oscillatory, unstable and unfair [174] especially towards connections with high round trip delays. Moreover, in networks incorporating wireless and satellite links, the long delays experienced, along with the increased noncongestion related losses, TCP protocols may also lead to network underutilization [45].

The above-mentioned problems lead to the development of congestion control designs based on solid mathematical formulations of network topologies.

In this chapter we shall focus on the fluid flow model formulation, whose study originates in [117] as well as in the seminal work [159].

6.2 The Fluid Flow Model

Information flow in packet switched networks is performed in a discrete fashion (i.e., an amount of packets are transferred in a specific time interval), which complicates congestion control analysis.

In fluid flow approximation, packets are assumed to flow continuously like fluids and be governed by nonlinear differential equations. In other words, two levels of abstraction are distinguished. The macroscopic (fluid-level) and the microscopic (packet-level). In the macroscopic level, congestion controllers are designed to achieve high utilization, low queueing delay - loss as well as fairness and stability. At the microscopic level, the aforementioned macroscopically designed algorithms are implemented to fulfill the end to end congestion control restrictions.

The reason behind the introduction of the fluid flow modeling is that using this setup the dynamic models derived are mathematically tractable, and are of sufficient accuracy. Apparently, fluid flow approximation error diminishes as the number of packets approaches infinity, or equivalently, as the packet size tends to zero.

Example 6.1 *(TCP-Reno dynamics) Reno is the congestion control algorithm in the current TCP that implements the standard AIMD policy, according to which the sending rate is regulated through the window size w. It is increased by $1/w$ packets every ACK and reduced to the half when a congestion event is detected. Let q denote the end to end packet loss probability and RTT the round trip time. Assume also that all packets are acknowledged. As a full window of packets is transmitted each RTT, on the average, the sending rate is*

$$x(t) = \frac{w(t)}{RTT(t)}.$$

To continue, on the average $1 - q(t)$ of the received ACKs are positive, each increasing the window by $1/w$. Therefore, the average window increase rate is

$$\frac{x(t)(1 - q(t))}{w(t)} = \frac{1 - q(t)}{RTT(t)}.$$

Similarly, the average window decrease rate is

$$\frac{x(t)q(t)w(t)}{2} = \frac{q(t)w^2(t)}{2RTT(t)}.$$

Hence, we conclude that on average, the window size evolves according to the fluid-flow model:

$$\dot{w} = \frac{1-q}{RTT} - \frac{qw^2}{2RTT}. \tag{6.1}$$

Setting $\dot{w} = 0$ in (6.1) we obtain the relation between packet loss probability and window size in equilibrium:

$$w^* = \sqrt{\frac{2(1-q^*)}{q^*}}. \tag{6.2}$$

When q^* is very small, (6.2) yields:

$$x^* = \frac{\sqrt{2}}{RTT} \frac{1}{\sqrt{q^*}} \tag{6.3}$$

which is the well-known $1/\sqrt{q^*}$ formula for TCP-Reno, first discovered in [174].

6.3 Network Representation

Let us consider a system of communication links indexed by $l = 1, 2, ..., L$, shared by a set of source-destination pairs indexed by $i = 1, 2, ..., N$. Each source i uses a set $\mathcal{L}_i \subseteq \{1, 2, ..., L\}$ of links. These sets define a $L \times N$ routing matrix R with elements:

$$R_{li} = \begin{cases} 1, & \text{if source } i \text{ uses link } l \\ 0, & \text{otherwise} \end{cases} \tag{6.4}$$

Each source i has an associated transmission rate $x_i(t)$ and let RTT_i be its corresponding round trip time.

Example 6.2 *To better understand the network topology representation, let us consider the network configuration displayed in Figure 6.1, which consists of 4 links, 5 sources and 4 flows.*

The i-th column of the routing matrix R reveals the path of the corresponding flow, while the l-th row of R informs about which flows the l-th link

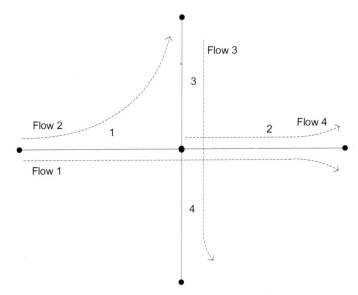

FIGURE 6.1
The network configuration of Example 6.2.

traverses; thus leading to the following routing matrix

$$R = \begin{bmatrix} 1 & 1 & 0 & 0 \\ 1 & 0 & 0 & 1 \\ 0 & 1 & 1 & 0 \\ 0 & 0 & 1 & 0 \end{bmatrix}.$$

The set of transmission rates determines the aggregate flow $y_l(t)$ at each link l by:

$$y_l(t) = \sum_{i=1}^{N} R_{li} x_i(t), \quad l = 1, 2, ..., L \tag{6.5}$$

where in (6.5) we have ignored the transmission delays from sources to links for simplicity. In vector form, (6.5) is expressed as

$$y(t) = Rx(t) \tag{6.6}$$

with $x = [x_1 \ x_2 \ ... \ x_N]^T$ and $y = [y_1 \ y_2 \ ... \ y_L]^T$. In (6.6) the routing matrix R was considered constant. However, generally, R is time varying. This is true owing to routing changes, as well as to users (sources) entering or leaving the network randomly. In such cases the constant routing matrix assumption is valid by considering that these variations happen at a slower time-scale than our analysis.

To each link a price $p_l(t)$ is associated, representing a congestion measure. Details on measuring congestion may be found in Chapter 4. Note that the price measure is a general formulation. When a source utilizes a loss-based protocol, p_l corresponds to packet loss probability, in delay-based protocols, p_l corresponds to queueing delay etc.

Aggregating all link prices in a route and dropping the transportation delays for simplicity, the aggregate route price is obtained:

$$q_i(t) = \sum_{l=1}^{L} R_{li} p_l(t), \quad i = 1, 2, ..., N \tag{6.7}$$

or in vector notation:

$$q(t) = R^T p_l(t). \tag{6.8}$$

The sources adjust their transmission rates according to the aggregate route price they receive. Hence,

$$\dot{\psi}_i = F_i(\psi_i, q_i), \quad i = 1, 2, ..., N \tag{6.9}$$

$$x_i = G_i(\psi_i, q_i), \quad i = 1, 2, ..., N. \tag{6.10}$$

Similarly, the links modify their prices based on the aggregate flow rates they receive. Therefore,

$$\dot{\xi}_l = H_l(y_l, \xi_l), \quad l = 1, 2, ..., L \tag{6.11}$$

$$p_l = Q_l(y_l, \xi_l), \quad l = 1, 2, ..., L. \tag{6.12}$$

The system described by (6.9), (6.10) represents the source dynamics, while (6.11), (6.12) constitute the link dynamics.

A hard constraint in determining (6.9) - (6.12) is that they must be decentralized. Stated otherwise, every source-link has access only to its local information. The latter design constraint shall be referred to as the end-to-end principle. Figure 6.2 illustrates in block diagram form the network representation.

6.4 Congestion Control as a Resource Allocation Problem

To each source a utility function $U_i(x_i)$, $i = 1, 2, ..., N$ is associated, describing the degree of satisfaction of the i-source with the particular transmission rate x_i assignment. Let us further denote with c_l the capacity of link l and define the capacity vector $C = [c_1 \ c_2 \ ... \ c_L]^T$.

The congestion control objective is to design F, G, H, Q in (6.9) - (6.12) such that

$$\lim_{t \to \infty} x(t) = x^* \tag{6.13}$$

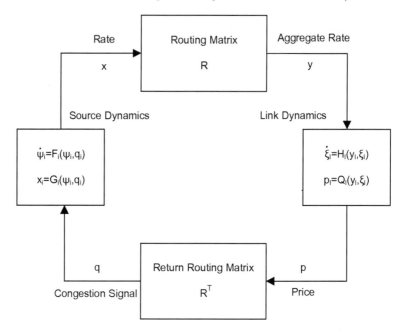

Source Dynamics

Link Dynamics

Congestion Signal

Price

FIGURE 6.2
Network flow control model.

with x^* being the solution of the following resource allocation problem:

$$\max_{\substack{Rx - C \leq 0 \\ -x \leq 0}} \sum_{i=1}^{N} U_i(x_i). \tag{6.14}$$

The capacity constraints $Rx - C \leq 0$ and non-negatively condition $x \geq 0$ defines a non-empty convex set. Furthermore, the utility functions $U_i(x_i)$, $i = 1, 2, ..., N$ are by construction strictly concave. Hence, (6.14) is a convex optimization problem. As such it has a unique optimum which can be found using Lagrangian methods [30], [26].

In the literature, the algorithms proposed to solve (6.14) are classified into:

a) primal [156], [276], [290], where the sources update their transmission rates using dynamic laws, while the links generate congestion signals using static expressions;

b) dual [70], [119], [178], [235], [282], where the links update the congestion signals dynamically, while the sources produce their transmission rates through static laws;

c) primal-dual [23], [69], [199], [265], [266], [291], where dynamic laws are utilized at both the sources and the links to determine the transmission rates and the congestion signals, respectively.

6.4.1 Dual Approach

The major issue in solving directly the problem (6.14) is that the coupling of the source transmission rates owing to the shared links through the capacity constraints leads to centralized solutions (i.e., sources should communicate information regarding their transmission rates), thus violating the end-to-end principle.

To overcome the above-mentioned issue, consider the Lagrangian function of (6.14):

$$L(x,p) = \sum_{i=1}^{N} U_i(x_i) - p^T(Rx - C). \tag{6.15}$$

Notice that in (6.15) the p vector represents the Lagrange multipliers and $q = R^T p$ is its aggregation.

Elaborating more, (6.15) yields

$$
\begin{aligned}
L(x,p) &= \sum_{i=1}^{N} U_i(x_i) - x^T R^T p + p^T C \\
&= \sum_{i=1}^{N} U_i(x_i) - x^T q + p^T C \\
&= \sum_{i=1}^{N} (U_i(x_i) - x_i q_i) + \sum_{j=1}^{L} p_j c_j.
\end{aligned}
\tag{6.16}
$$

With the help of L, the dual problem of (6.14) is defined as follows:

$$\min_{p \geq 0} \max_{x \geq 0} L(x, p). \tag{6.17}$$

The convex duality property implies that the optimum (x^*, p^*) of (6.17) also solves (6.14).

Following the analysis of [136] and after a full row rank assumption for the routing matrix R, it can be proven that the system of differential equations

$$
\dot{p}_j = \begin{cases}
y_j - c_j, & \text{if } p_j > 0 \text{ or if } p_j = 0 \text{ and } y_j - c_j > 0 \\
0, & \text{if } p_j = 0 \text{ and } y_j - c_j \leq 0
\end{cases}
\tag{6.18}
$$

$$p_j(0) \geq 0 \tag{6.19}$$

$$x_i = \left[\frac{dU_i}{dx_i(q_i)} \right]^{-1} \tag{6.20}$$

for all $i = 1, 2, ..., N$ and $j = 1, 2, ..., L$, has a unique equilibrium p^*, which is asymptotically stable and the corresponding vector x^* is the unique solution to the problem (6.14).

6.4.2 Primal Approach

The dual algorithm (6.18), (6.20) is decentralized indeed as it uses only local information to solve the problem (6.14). However, all decisions are taken on the link side of the network and not on the "end-users" which are the sources, thus destroying the fundamentals of the Internet deployment.

The primal approach was motivated by the necessity of developing algorithms operated at the source side of the network, always fulfilling the end-to-end principle. In that respect, an approximation to the problem (6.14) is provided, by considering the following optimization problem:

$$\max_{x \geq 0} \left[\sum_{i=1}^{N} U_i(x_i) - \sum_{j=1}^{L} \int_0^{y_j} R_{ji} f_j(\tau) d\tau \right] \tag{6.21}$$

where the functions f_j, $j = 1, 2, ..., L$ are selected such that

$$V(x) = \sum_{i=1}^{N} U_i(x_i) - \sum_{j=1}^{L} \int_0^{y_j} R_{ji} f_j(\tau) d\tau$$

is coercive (i.e., $\lim_{|x| \to \infty} V(x) = -\infty$). In other terms, the prices are treated as penalty functions for the capacity constraints.

It has been shown in [136] that the system of differential equations

$$\dot{x}_i = \begin{cases} \frac{dU_i}{dx_i}(x_i) - q_i, & \text{if } x_i > 0 \text{ or if } x_i = 0 \text{ and } \frac{dU_i}{dx_i}(x_i) - q_i > 0 \\ \\ 0, & \text{if } x_i = 0 \text{ and } \frac{dU_i}{dx_i}(x_i) - q_i \leq 0 \end{cases} \tag{6.22}$$

$$q_i = \sum_{j=1}^{L} R_{ji} f_j(y_j) \tag{6.23}$$

with y_j as defined in (6.5) and $x_i(0) \geq 0$, $i = 1, 2, ..., N$ has a unique equilibrium point x^*, which is globally asymptotically stable and is the unique solution of the problem (6.21).

6.4.3 Utility Function Selection

The analysis presented thus far clearly reveals the strong bond of the solution of the congestion control issue, formulated as a resource allocation problem, with the utility function. In this subsection we provide a number of design considerations.

Application Oriented

Current Internet pervades a large number of applications with diverse service requirements, a fact that is expected to become even more severe in the near future. For example applications like VoIP require the availability of a constant bandwidth; others are sensitive to jitter; whereas others can occasionally tolerate more effectively packet losses.

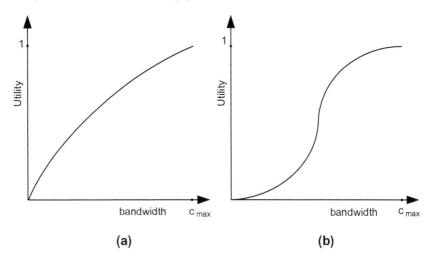

(a)   **(b)**

FIGURE 6.3
Utility curves for (a) elastic applications; (b) adaptive real-time traffic applications.

The effectiveness of an application is evaluated through a satisfaction metric, which can be represented by a utility function that maps network characteristics like bandwidth, queueing delay, jitter and packet loss, to application features. Different utility functions may be attached to different applications. It has been reported [294] that the utility surface for elastic applications like SMTP, FTP, HTTP etc. is furnished by a strictly non-decreasing concave function, as pictured in Figure 6.3(a), while delay-adaptive applications[1] have an S-shaped utility surface (see Figure 6.3(b)). To simplify the representation, in Figure 6.3 we have considered single-argument utility functions.

Fairness Considerations
It has been shown that the choice of the utility function strongly determines the equilibrium properties of a primal and/or a dual congestion control protocol. For the class of concave utility functions parameterized by $a_i \geq 0$ and $p_i > 0$

$$U_i(x_i) = \begin{cases} p_i \ln x_i, & a_i = 1 \\ p_i(1 - a_i)^{-1} x_i^{1-a_i}, & a_i \neq 1 \end{cases} \tag{6.24}$$

the congestion control algorithm obtained is $(p_i - a_i)$-proportionally fair [262].

---

[1]Applications that dynamically adapt to changing packet transmission delays and to dynamically change the level of packet delivery delay control they request from the network when their current level of service is not adequate.

The aforementioned fairness results hold in a homogeneous environment. Heterogeneity (i.e., different protocols based on different pricing schemes that share the same network) introduces multiple equilibria, complicating significantly the analysis [210].

6.5 Open Issues

Despite the significant advances in Internet congestion control, still many issues persist raising burdens on the wide deployment of new types of applications that require high quality in both the network and the application levels. In this section the most significant open problems will be summarized and explained.

6.5.1 Stability and Convergence

The results presented in the previous section hold under a stability assumption. Stated otherwise, it was implicitly assumed that the primal or the dual algorithms were capable of driving the system to its equilibrium.

Stability is a very important issue [153], [281], [204], [66], [186], [8]. An unstable protocol introduces large fluctuations in the queue lengths, which in turn reduce utilization, throughput and increase jitter, degrading performance, especially of applications that require interactive services such as VoIP.

Needless to mention, the equilibrium, even in fully homogeneous network environments, typically varies with time, as the number of users (i.e., sources) entering or leaving the network also varies with time. It is therefore desirable for the network to adapt efficiently (sufficiently fast and smooth) to the changing operational conditions. In that perspective, understanding the stability and convergence properties of Internet congestion control algorithms is imperative.

Control theory is the natural framework for studying stability and convergence issues for dynamical systems like the Internet. Besides achieving asymptotic convergence to an equilibrium point, the Internet congestion control may exhibit controlled oscillation points (i.e., limited cycles with bounded variations). In both cases the objective of a congestion control algorithm is to achieve steady-state performance in terms of link utilization, throughput and round trip time.

Control theoretic approaches implemented as congestion controllers in the links have been reported [163], addressing local stability around an equilibrium, since linearized versions of the fluid flow model and queue dynamics were considered.

Primal algorithms have also been reported [159], [193], [194] leaving, however, certain key questions open for future investigation. Specifically:

a) a model for the controlled system (network) has to be carefully devised, otherwise stability analysis is impossible;

b) the determination of a desired and feasible system performance is a nontrivial task owing to the highly uncertain and dynamically varying nature of the Internet;

c) the output of any proposed control scheme should be saturated to avoid the violation of the link capacity constraints of the network.

The speed of convergence (responsiveness) of a congestion control algorithm to a stable equilibrium determines, to a great extent, its overall performance. The significance of this metric enlarges as the network bandwidth - delay product increases (i.e., long-distance high-speed networks).

6.5.2   Implementation Constraints

Any congestion control algorithm should be decentralized (i.e., should have access only to local information) and no exchanging of information should take place among different control sites.

A candidate dual algorithm should be simple. In this way the total number of computations performed owing to congestion control in a router of a network is kept to a minimum, a property of great significance especially when high-speed networks are considered.

Real-time constraints are imposed on primal algorithms as well. However, as these are located at the endpoints (sources) of a network, fairly complex implementations are allowed.

The Internet was deployed and is currently expanding on the understanding that each user (source) regulates its transmission rate based on feedback received from the network, which provides congestion notification measures without having access to link algorithms, or knowing any kind of information regarding link-network characteristics (degree of heterogeneity, the presence of wireless or satellite links, the bandwidth-delay product of the network etc.), or even to assumptions on network traffic dynamics.

The aforementioned practical restrictions establish primal algorithms as the only congestion control scheme that fulfills the end-to-end network considerations.

6.5.3   Robustness

An Internet congestion control scheme should not only be stable but additionally it should be robustly stable against network modeling imperfections and exogenous disturbances.

All sources in a network do not follow the same congestion control algorithm. For example UDP applications introduce unresponsive traffic, which appears as a disturbance to the dominating TCP congestion control schemes.

If the congestion controller is not robust enough to compensate for their effects, the network may even de driven to congestion collapse [139].

Besides the above, many sources of uncertainties exist in the Internet, which are mainly caused by the inherent heterogeneity, the large variation in network operating conditions (link speeds, delay, link buffer capacity, link type etc.), as well as errors in measuring the end-to-end congestion level from the received aggregate congestion notification signals (typically of binary type) issued by the network. A robust congestion control algorithm should be insensitive to the various traffic environments.

Nevertheless, developing a new congestion controller or improving an existing one has to be done without affecting the robustness of the congestion control algorithms operating at the network. To deal with modeling uncertainties, the recently developed systems discipline of approximation-based control (i.e., neural network control, fuzzy control etc.) that was originally proposed for on-line black box system identification and robust adaptive control of highly uncertain and nonlinear dynamical systems, has now reached a maturity level, making it a promising working framework for developing source implemented (primal) congestion controllers.

6.5.4   Fairness

Fairness implies fair bandwidth among competing sources. A bandwidth allocation scheme is fair if it treats all connections in the same manner, either based on the time order in which they ask for the available bandwidth, or the location of their endpoints.

A common assumption behind fair sharing is that all sources are operating under the same congestion control algorithm (intrafairness). The fairness issue has become increasingly important owing to the appearance of long-distance high-speed networks that incorporate large bandwidth-delay products, together with the user-demand for high bandwidth and low delay communications. The presence of unresponsive network traffic (e.g., UDP) complicates the fairness issue even more.

Heterogeneity has introduced a different fairness concept, called interfairness, that studies fair bandwidth sharing among competing heterogeneous sources (implement conceptually different congestion control schemes). Owing to the high complexity and the difficulty of the problem, reported interfairness studies are constrained to simulation studies [150], [179]. Rigorous analysis is currently an open issue.

6.6 Concluding Comments

In this chapter an introduction has been provided to the congestion fluid flow model of the Internet, which is used to provide macroscopic understanding of congestion control schemes. After formulating it as a resource allocation problem, primal and dual congestion controllers were derived. Open issues related to stability, convergence, robustness, fairness as well as implementation constraints were clearly identified, indicating the weak points that require special attention and treatment. In the next chapter a new congestion control framework will be formulated to provide alternative solutions, at least to a significant class of the aforementioned issues by exploiting principles borrowed from robust adaptive control, shown to naturally apply to this type of problem.

Part II

Adaptive Congestion Control Framework

7

NNRC: An Adaptive Congestion Control Framework

CONTENTS

7.1   Overview

This book focuses on unicast single rate congestion control mechanisms able to support flow control at sources which are entrusted with real-time applications. The main purpose of a congestion control mechanism is to achieve high network utilization, small queueing delays and some degree of fairness among flows. Furthermore, the introduction of new types of services in the Internet enforced new quality requirements (Quality of Service) at congestion control mechanisms that should be fulfilled even in congested network conditions.

Furthermore, any candidate congestion control mechanism should be still operational despite the presence of long-distance high-speed and/or wireless connections.

In this chapter, an adaptive congestion control framework (*NNRC*) is presented and its main operational characteristics and functionalities are qualitatively analyzed. Dominantly, *NNRC* regulates the per packet round trip time to follow accurately and reliably a reference signal, while preventing link buffers from overflow (congestion), as well as empty queues (poor utilization). Furthermore, the aforementioned operation is achieved in a fair manner, meaning that all network resources are equally allocated among competing sources.

After stating the problem, the *NNRC* framework is briefly presented. It can be seen that it is comprised of four interconnected modules. Specifically, of i) a future path congestion level estimator; ii) an on-line feasible desired round trip estimator; iii) an adaptive and saturated transmission rate controller and iv) a throughput controller.

The most significant attribute of *NNRC* is probably its learning ability; allowing the protocol ingredients to dynamically adapt to changing network conditions, while maintaining stability and performance.

Rigorous theoretical analysis of the *NNRC* framework will be provided in Chapters 8 and 9, while extensive simulation tests and comparisons appear in Chapter 10.

7.2 Packet Switching Network System

We consider a general packet switching network (see Figure 7.1), whose topology is characterized by a set of sources/receivers $C= \{1, 2, ..., n\}$, a set of nodes[1] $Q= \{1, 2, ..., m\}$ and a set of links $\mathcal{L}= \{1, 2, ..., l\}$ connecting the nodes. Each link $l \in \mathcal{L}$ has an associated buffer with maximum capacity B_l. A source $S \in C$ has to transmit an application of a prespecified amount of packets N to a destination $D \in C$ through the network with transmission rate ($u \in (0, \overline{u}]$) controlled by an appropriately designed protocol. The transmission rate is the frequency the source places packets on its output buffer. Every source - receiver pair is characterized by a path $L(S, D)$, defined as the set of links the transmitted packets follow to reach their destination.

Upon arrival of a packet, the destination (receiver) issues an acknowledgment (ACK), which is received by the source. The amount of time that elapses between the instant the source starts to transmit a packet and the instant at which it receives its ACK is called round trip time (RTT). In case of a packet loss (e.g., owing to a buffer overflow inside the network), no ACK will be received. The source waits for a certain amount of time called the timeout interval (τ_0), after which, if no ACK is received, the source proceeds to the packet retransmission.

Assuming for a moment that a single packet is transmitted every round trip time and no packet loss is present, sending time[2] T_s may be calculated on the basis of RTT as:

$$T_s = \int_0^N RTT(k)dk. \tag{7.1}$$

The round trip time (RTT) of a source - receiver connection via a path $L(S, D)$ is comprised of the transmission time T_t (defined as $T_t = p_{size}/u$, $u \in (0, \overline{u}]$ with $p_{size}, \overline{u}$ denoting the packet size and the maximum transmission rate, respectively), the path propagation delay d_p, as well as the buffering delay

---

[1] We use the terms node and router interchangeably in this book.

[2] The amount of time required for transmitting the application.

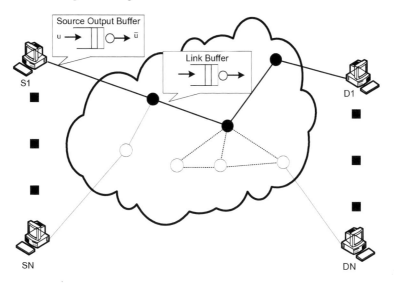

FIGURE 7.1
The packet switching network topology.

d_b formed in every non-empty buffer in $L(S, D)$. In other terms:

$$d_p = \sum_{j \in L(S,D)} d_p(j) \qquad (7.2)$$

$$d_b = \sum_{j \in L(S,D)} d_b(j) \qquad (7.3)$$

where $d_p(j)$ is the propagation delay of the j-th link and $d_b(j)$ is the packet delay experienced in the j-th buffer (see Figure 7.2). It is harmless to consider constant propagation delay d_p when $L(S, D)$ is comprised of only wired connections. However, such a consideration is not valid in case at least one wireless connection is present.

If the incoming flow rate to a link exceeds its bandwidth, packets are queued in the link buffer and the long duration of this incident leads to congestion collapse (buffer overflow). Each link estimates a measure of its congestion level (denoted by p_j, $j \in L(S, D)$), which depends on the value of the incoming flow rate and the current buffer length. This information is transmitted through the path $L(S, D)$ to the destination node, at which an aggregate path congestion measure p is formed via a nonlinear mechanism. In other terms

$$p = F(p_1, p_2, ..., p_l), p \in [0, 1] \qquad (7.4)$$

with l denoting the number of links in $L(S, D)$.

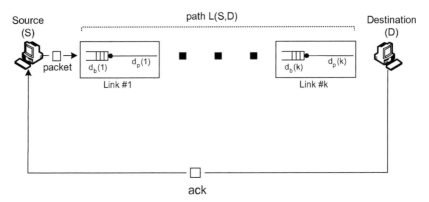

FIGURE 7.2
Transferring packets through the path $L(S, D)$.

Specifically, assuming [186], [194] that only the i-th link in a path is congested, then $p_i >> \sum_{j \neq i, j \in L(S,D)} p_j$ and hence with no loss of generality, (7.4) takes the form:

$$p = \sum_{l \in L(S,D)} p_l. \tag{7.5}$$

In the literature,[3] various approaches have been proposed to reliably convey coded path congestion information to the corresponding source. It is there where such information is decoded and consequently employed in the determination of the current transmission rate. In this book, whenever a packet arrives to be forwarded on a link, it is marked (set the ECN bit in the packet's header) with a probability whose value depends on the link buffer length. In this way, a clear connection between congestion and buffering delay is established.

Remark 7.1 *We have to underline that a zero path congestion measure (p) represents the almost empty buffers scenario. However, $p = 1$ does not necessarily mean that the network is already in congestion (buffer overflow). In reality, $p = 1$ reveals that the corresponding path $L(S, D)$ is approaching congestion. This is especially true whenever the links incorporate marking mechanisms for congestion notification.*

---

[3]For further details on developing coded path congestion information, the interested reader may consult Chapter 4 and the references therein.

7.3   Problem Statement

For the specific packet switching network system described in the previous subsection, our primary concern is to design decentralized rate controllers, implemented at the source side, capable of guaranteeing high sensitivity to time quality characteristics, such as almost constant per packet delay, while preventing the path $L(S, D)$ from congestion collapse and throughput from starvation.

The aforementioned design implies the existence of a control algorithm, capable of regulating the per packet RTT close to a desired round trip time RTT_d, avoiding either overflow or empty link buffers. Let us assume for a moment that the source is about to transmit the k-th packet. Having knowledge of the path congestion level p this packet will experience in its travel through $L(S, D)$, it is reasonable to claim that there exists a smooth, bounded but unknown nonlinear function of RTT, u, p such as:

$$\dot{RTT} = f(RTT, u, p), \quad RTT \in [0, \tau_0], \quad u \in (0, \overline{u}], \quad p \in [0, 1). \qquad (7.6)$$

Remark 7.2 *Owing to (7.6), RTT variations depend not only on past RTT values and the current transmission rate u, but most importantly on the future path congestion level p. In this respect, significant abrupt changes in p may be more effectively incorporated into the u design.*

Besides the uncertainty in $f()$, (7.6) also requires knowledge of the future path congestion level p, which is unfortunately unknown and thus has to be estimated. In addition, the a priori availability of feasible RTT_d values is not a straightforward task, owing to the highly uncertain and dynamically varying character of the packet switching network.

Hence, our control problem is re-formulated as follows:

Problem 7.1 *Design:*

- *a future path congestion level estimator,*

- *a feasible RTT_d estimator,*

- *a source rate controller*

capable of regulating $e = RTT - RTT_d$ to an arbitrarily small neighborhood of zero, while keeping all other signals in the closed-loop uniformly bounded. All algorithms should be decentralized and implemented at the source.

Remark 7.3 *If RTT_d is designed to be piecewise constant, then regulating e to an arbitrarily small neighborhood of zero, guarantees the almost constant per packet delay requirement, thus achieving the time quality characteristics.*

The proposed transmission control framework in block diagram form is pictured in Figure 7.3.

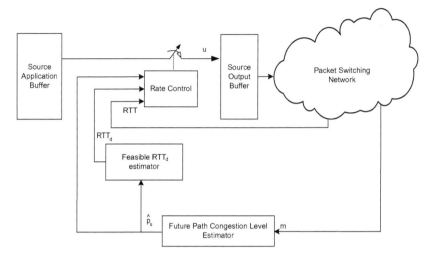

FIGURE 7.3
The proposed transmission control framework in block diagram form.

7.4   Throughput Improvement

In the proposed transmission control framework, each source transmits one packet every RTT and remains idle after transmission. However, the number of packets per source is high, which unavoidably leads to a significant increase in sending time (T_s) and throughput reduction. Apparently, improvement can be achieved by reducing the aforementioned idle time.

To alleviate this problem, the notion of communication channels is introduced. Each source creates a number (η) of communication channels that operate in parallel as if each one is a separate source.

Every channel draws packets from the common source application buffer and has its own transmission rate. Channels place the packets to the source output buffer which is now viewed as an extra link buffer in the path $L(S, D)$, and thus the corresponding delay is included in the total path buffering delay defined previously.

The aforementioned analysis leads to the addition of an extra module, named throughput control, which is responsible for the regulation of the number of channels. The general architecture of the proposed transmission control framework is presented in Figure 7.4.

Besides throughput improvement, the corresponding module attains the responsibility of improving fairness characteristics[4] on the operation of $NNRC$ framework.

---

[4]Fair sharing of bandwidth among competing sources.

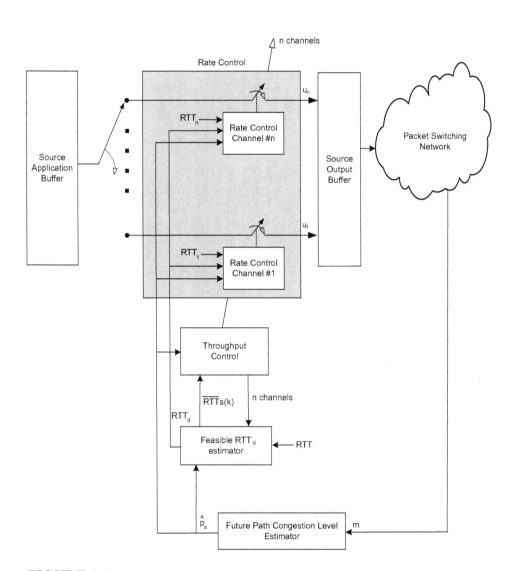

FIGURE 7.4
General architecture.

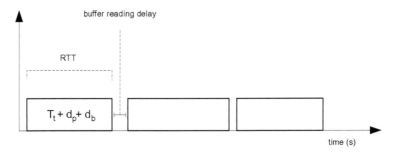

FIGURE 7.5
Buffer reading delay.

Channels Implementation

Each channel at a source is visualized as a process designed to forward packets from the common application buffer to the source output buffer. More specifically, the rate controller attached to every channel makes a call to the memory module (source application buffer) to receive the next packet for transmission. Such a sequential implementation has the advantage of reduced complexity, as no dedicated hardware is required, constituting implementation software oriented. However, it may also lead to the appearance of an extra delay, called buffer reading delay, defined as the time interval between the instant a channel makes a packet request to the memory module and the moment this request is granted. The phenomenon is pictured in Figure 7.5.

Buffer reading delay enlarges, as the number of channels that concurrently issue packet requests increases. Even though it cannot be theoretically avoided, extensive simulations have shown that the phenomenon mainly occurs at the initialization phase (where all channels request simultaneous transmission) thus significantly reducing any negative impact on the performance of the *NNRC* framework.

We are closing the discussion on channels implementation by stressing that the *NNRC* theoretical analysis that will be presented in the subsequent chapters assumes that all channels are connected in parallel to the source application buffer. Fortunately, the proposed sequential implementation introduces no side-effects to the analysis, as the buffer reading delay is not included in RTT calculation.

7.5 *NNRC* Framework Description

Continuing the discussion presented in the previous sections, we state below a generalized version of Problem 7.1 which also summarizes the desirable performance characteristics of the *NNRC* framework.

Problem 7.2 *Design:*

    1. a future path congestion level estimator;

    2. a feasible desired round trip time estimator;

    3. a throughput controller;

    4. a rate controller

capable of guaranteeing packet round trip time close to a desired value (RTT$_d$), fair sharing of bandwidth among competing sources, while preventing the path from congestion collapse and throughput from starvation.

In what follows the *NNRC* framework modules shall be qualitatively presented emphasizing their operational characteristics. Detailed mathematical analysis and verification are the main themes of Chapters 8 and 9.

7.5.1 Future Path Congestion Level Estimator

Each packet, when travelling through $L(S, D)$, experiences an aggregate path congestion measure $p(k)$, $k = 1, 2, ..., N$ and every source requires knowledge of the future path congestion level $p(k+1)$ to determine the transmission rate. Therefore, the need for constructing on every source a future path congestion estimator is evident. In that respect a one-bit on-line assignment scheme is proposed. Specifically, consider receiving the ACK of the k-th packet. The incoming ECN bit $m(k)$ is either 0 or 1 depending on whether the packet is marked or not. Let $\hat{p}_s$ denote the estimate of p. The below-mentioned recursive formula (stable low-pass filter) is employed for the derivation of $\hat{p}_s(k + 1)$:

$$\hat{p}_s(k+1) = \alpha\hat{p}_s(k) + (\alpha-1)m(k), \quad k = 1, 2, ...N, \quad \hat{p}_s(0) = 0, \quad 0 < \alpha < 1 \quad (7.7)$$

where $\hat{p}_s(0)$ denotes the initial estimate of p. The design constant α controls the significance of old estimates in deriving the new one. From (7.7) and the definition of $m(k)$ it becomes apparent that $\hat{p}_s \in [0, 1]$.

Remark 7.4 *There are two sources of error in the above-described estimation process. The first, which has the form of modeling error, is introduced when selecting (7.7) as a means to derive $\hat{p}_s$. Apparently, (7.7) constitutes the simplest possible implementation one can employ to obtain future path congestion estimated values. Certainly, other more complicated structures (i.e., neural networks, fuzzy systems) may be used to attenuate this source of error, raising however significantly the complexity. Furthermore, owing to significant delays (propagation and buffering) in the path $L(S, D)$, the estimates $\hat{p}_s$ as obtained from (7.7), actually correspond to past p value. The latter constitutes a major source of error in $\hat{p}_s$ derivation.*

Owing to Remark 7.4 we may write:

$$p = \hat{p}_s + \tilde{p} \quad (7.8)$$

with $\tilde{p}$ denoting the total estimation error. The presence of a non-zero $\tilde{p}$ affects the stability analysis of the closed-loop system. Detailed consideration will be provided in Chapters 8 and 9.

7.5.2   Feasible Desired Round Trip Time Estimator

NNRC framework requires knowledge of a desired and feasible round trip time per packet value (RTT_d) for all application packets. Unfortunately, owing to the highly uncertain and dynamic nature of the Internet, no such information is available to the source a priori. Therefore, the design of a corresponding estimation scheme is mandatory. Within the proposed framework such an estimation module accepts as inputs current and past RTT and $\hat{p}_s$ values and outputs a feasible desired round trip time (RTT_d) signal as well as the group mean round trip time $\overline{RTT}_s(k)$. The latter acts as a reference at the rate control unit.

It is to be noted however that the feasible RTT_d estimation module will not be designed to search for the minimum RTT per packet that can be achieved in the existing network conditions (i.e., congestion level). In fact, such an optimization process could eventually lead to empty link buffers and consequently to throughput starvation. Stated otherwise, feasible RTT_d is not equivalent to minimum RTT_d.

$$\boxed{\text{Feasible } RTT_d \neq \text{Minimum } RTT_d}$$

The aforementioned feasibility discussion is given network - wise. Constraints on the feasible RTT_d may also be imposed by closely examining the application specifications, establishing in this way application - wise feasible RTT_d considerations. Even though combining network - wise with application - wise feasibility considerations is not a trivial task, the proposed *NNRC* framework could serve as a vehicle for achieving this purpose. To maintain generality, the *NNRC* theoretical analysis presented in Chapters 8 and 9 will deal solely with network - wise feasibility.

7.5.3   Rate Control

Receiving as inputs a) the number of channels, b) a feasible RTT_d, c) an estimate of the future path congestion level $\hat{p}_s$, and d) the current RTT measurements, the rate control unit regulates at each communication channel the packet transmission time T_t, to reliably and accurately follow the reference signal RTT_d downloaded from the feasible RTT_d estimator.

An important operational functionality of this module concerns designing channel rate controllers to increase the corresponding packet transmission times in case of strong congestion evidence in the path ($\hat{p}_s = 1$). As will be clarified in Chapters 8 and 9, the $\hat{p}_s = 1$ scenario is equivalent to an uncontrollable situation. In that respect, reducing the overall source transmission rate is the only acceptable alternative.

7.5.4 Throughput Control

As argued in Section 1.4, throughput within *NNRC* can be substantially improved by incorporating the notion of communication channels, whose regulation is the main responsibility of the throughput control unit. The controller designed for that purpose receives as inputs current p estimates produced by the future path congestion level estimator, as well as the group mean round trip time $\overline{RTT}_s(k)$ which is downloaded from the feasible desired round trip time estimation unit and adapts the actual number of channels implemented to the rate control unit.

As will be clarified in Chapter 8, throughput control operates on a different time-scale compared to rate control, with the latter representing the fast dynamics. In that respect the number of channels used remains constant throughout rate adaptation.

Besides improving throughput, this unit is also responsible for achieving fair sharing of bandwidth among competing sources. An interesting positive side effect concerns improving *NNRC* reactions to sudden unpredictable network variations e.g., re-routings, bursty traffic conditions etc. The latter operational characteristics will be verified in Chapters 9.

7.6 Concluding Comments

In this chapter we have presented, in a qualitative manner, an adaptive congestion control framework (*NNRC*), focusing on its main operational characteristics and functionalities.

According to the *NNRC* philosophy, regulation of the per packet round trip time is essential to achieve accurately and reliably time-quality application oriented constraints, while preventing link buffers from overflow (congestion) or empty queues (underutilization), under fairness considerations.

NNRC is the first adaptive congestion control framework presented in the relevant literature designed to operate at the source. Early versions of *NNRC* can be traced back at 2005 (see [125], [126], [127], [128], [129]). The final version was formed with the aid of [130], [131]. Among its key operational advantages are adapticity and modularity In fact each one of the four modules plays a specific role within *NNRC*. As a consequence, different designs can be employed and implemented inside the modules as long as its key functionalities are maintained. This is why the word "framework" is adopted to characterize *NNRC* instead of "protocol." Each different design leads to a different protocol under the umbrella of *NNRC*.

8

NNRC: *Rate Control Design*

CONTENTS

8.1   Overview

This chapter focuses on the design of a) the feasible round trip time estimator and b) the rate control module of the *NNRC* framework. Throughout the chapter the number of channels used is constant to a value known in advance. The latter assumption will be relaxed in Chapter 9 when considering the throughput control unit. Robustness against modeling imperfections, exogenous disturbances (UDP traffic) and delays (propagation and buffering) are proven. The developed rate is guaranteed to be saturated, while modifications are provided to achieve its reduction whenever congestion is detected. Simulations performed on a small-scale network highlight the main attributes of the designed *NNRC* modules.

8.2   Feasible Desired Round Trip Time Estimator Design

In the proposed transmission control framework, each source has to transmit an application of N packets to a destination through the network. The adaptive control formulation requires knowledge of a desired *RTT* per packet value (RTT_d) for all application packets. Unfortunately, no such information is available at the source a priori. To establish a feasible solution, an estimated sending time mechanism should be constructed at the source, aiming at providing reliable reference to the rate controller.

Within the *NNRC* framework, each application is divided into groups of M packets and the channels undertake the parallel transmission of the packets in the group. Let us assume that we have already transmitted k groups and we are about to transmit the $k + 1$ group. Let $\overline{RTT}_s(k + 1)$ and $\overline{RTT}_s(k)$ be the mean round trip times for the $k + 1$ and the k groups, respectively, while $\hat{p}_s(k)$ denotes the path congestion level estimated via (7.2), at the beginning of the k-th group transmission.

Assuming for the moment almost constant $\hat{p}_s$ during the transmission of the k-th group, we argue that the variation $\Delta \overline{RTT}_s(k) = \overline{RTT}_s(k + 1) - \overline{RTT}_s(k)$ depends on past path congestion levels and mean round trip times. Hence, $\Delta \overline{RTT}_s(k)$ can be described by:

$$\Delta \overline{RTT}_s(k) = \Phi(D(k)), \quad \overline{RTT}_s(0) = 0 \tag{8.1}$$

where $\Phi(D(k))$ is an unknown, sufficiently smooth and bounded function of $D(k) = \{\overline{RTT}_s(k), \overline{RTT}_s(k-1), \overline{RTT}_s(k-2), ..., \overline{RTT}_s(k-m_1), \hat{p}_s(k), \hat{p}_s(k-1), \hat{p}_s(k-2), ..., \hat{p}_s(k-m_2)\}$ with properly selected positive integers m_1, m_2.

To continue, owing to the approximation capabilities of the linear in the weights neural networks, we can assume, with no loss of generality, that there exists weight values $W_s^* \in \Re^{L_s}$ and an appropriately defined regressor vector $S_s \in \Re^{L_s}$, such that the unknown system (8.1) can be completely described by

$$\overline{RTT}_s(k + 1) = \overline{RTT}_s(k) + W_s^{*T} S_s(D(k)) + \omega_s(D(k)). \tag{8.2}$$

Owing to the neural networks density property, there exists an arbitrarily small positive constant δ_s such that $|\omega_s(D(k))| \leq \delta_s$, $\forall \overline{RTT}_s \in \Omega_s$ and $\forall \hat{p}_s(k) \in [0, 1)$, where Ω_s is a compact region.

Defining $\hat{W}_s(k)$ as the estimate of W_s^* at group k, we generate the estimated value $\widehat{\overline{RTT}}_s(k + 1)$ of the mean round trip time $\overline{RTT}_s(k + 1)$ as

$$\widehat{\overline{RTT}}_s(k + 1) = \widehat{\overline{RTT}}_s(k) + \hat{W}_s^T(k) S_s(D(k)) + \beta e_s(k) \tag{8.3}$$

where $\beta \neq 0$ and $e_s(k)$ is the estimation error defined as

$$e_s(k) = \overline{RTT}_s(k) - \widehat{\overline{RTT}}_s(k). \tag{8.4}$$

From (8.2), (8.3) we obtain:

$$e_s(k + 1) = (1 - \beta)e(k) - (\hat{W}_s(k) - W_s^*)^T S_s(D(k)) + \omega_s(D(k)) \tag{8.5}$$

where $\beta \neq 0, 1$.

Define the weight estimation error as

$$\tilde{W}(k) = \hat{W}_s(k) - W_s^*. \tag{8.6}$$

To derive stable control and update laws, Lyapunov stability theory is employed. Let us take the Lyapunov function candidate

$$J = e_s^2(k) + \hat{W}_s^T(k)\hat{W}_s(k). \tag{8.7}$$

Its first difference is given by

$$\Delta J = e_s^2(k+1) - e_s^2(k) + \hat{W}_s(k+1)^T\hat{W}_s(k+1) - \hat{W}_s^T(k)\hat{W}_s(k). \tag{8.8}$$

Take the weight update law as

$$\hat{W}_s(k+1) = \hat{W}_s(k) + \mathcal{P}_s\left\{(1-\beta)e_s(k)S_s(D(k))\right\} \tag{8.9}$$

where $\mathcal{P}_s$ denotes the projection operator with respect to the convex and bounded set

$$\mathcal{R}_s = \{\hat{W}_s \in \Re^{L_s} : \|\hat{W}_s\| \le w_s, \ w_s > 0\}. \tag{8.10}$$

In (8.9) the projection operator is defined as

$$\mathcal{P}_s\{z\} = \begin{cases} z & \text{if } \|\hat{W}_s(k) + z\| \le w_s \\ z - z\left(1 - \frac{w_s - \|\hat{W}_s(k)\|}{\|z\|}\right) & \begin{array}{l}\text{if } \|\hat{W}_s(k) + z\| > w_s \\ \text{and } z\hat{W}_s(k) \ge 0\end{array} \end{cases}. \tag{8.11}$$

Lemma 8.1 *If the initial weights are chosen such that $\hat{W}_s(0) \in \mathcal{R}_s$ and $W_s^* \in \mathcal{R}_s$ then $\hat{W}_s(k) \in \mathcal{R}_s$ for all $k \ge 0$.*

Lemma 8.2 *Based on the projection modification, the additional terms introduced in the expression for J, can only make ΔJ more negative.*

The proofs of Lemmas 8.1, 8.2 are provided in the subsections that follow. Substituting (8.3)-(8.6) and (8.9) in (8.8) we finally obtain:

$$\begin{aligned}\Delta J = & \left[\beta(\beta - 2) + (1-\beta)^2|S_s(D(k))|^2\right]e_s^2(k) \\ & + 2(1-\beta)\left[w_s(D(k)) + W_s^{*T}(k)S_s(D(k))\right]e_s(k) \\ & + |\tilde{W}_s^T(k)S_s(D(k))|^2 + \omega_s^2(D(k)) \\ & - 2\omega_s(D(k))\tilde{W}_s^T(k)S_s(D(k)).\end{aligned}$$

Since $S_s(D(k))$ is bounded by construction and $\hat{W}_s(k)$ is bounded owing to Lemma 8.1, we have $|S_s(D(k))| \le \overline{S}_s$ and $|\tilde{W}_s(k)| \le 2w_s$. Thus,

$$\begin{aligned}\Delta J \le & \ \bar{\beta}|e_s(k)|^2 + 2|1 - \beta|(\delta_s + w_s\overline{S}_s)|e_s(k)| \\ & + 4|\tilde{W}_s^T(k)|^2\overline{S}_s^2 + \delta_s^2 + 2\delta_s w_s\overline{S}_s\end{aligned}$$

where $\bar{\beta} = (1 + \overline{S}_s^2)\beta^2 - 2(1 + \overline{S}_s^2)\beta + \overline{S}_s^2$ with $\bar{\beta} < 0$ whenever

$$\beta \in \left(1 - \sqrt{1/(1 + \overline{S}_s^2)}, 1\right) \bigcup \left(1, 1 + \sqrt{1/(1 + \overline{S}_s^2)}\right).$$

Hence, $\Delta J \le 0$ provided that:

$$|e_s(k)| \ge \frac{-|1-\beta|(\delta_s + w_s \overline{S}_s)}{\bar{\beta}} + \sqrt{\frac{(\delta_s + w_s \overline{S}_s)^2|1-\beta|^2}{\bar{\beta}^2} - \frac{\left[|\tilde{W}_s^T(k)|^2 \overline{S}_s^2 + \delta_s^2 + 2\delta_s w_s \overline{S}_s\right]}{\bar{\beta}}}.$$

Therefore, $e_s(k)$ possesses a uniform ultimate boundedness property with respect to the set:

$$\begin{aligned}\mathcal{E}_s \;=\; &\left\{ e_s(k) : |e_s(k)| < \frac{-|1-\beta|(\delta_s + w_s \overline{S}_s)}{\bar{\beta}}\right.\\ &\left.+\sqrt{\frac{(\delta_s + w_s \overline{S}_s)^2|1-\beta|^2}{\bar{\beta}^2} - \frac{\left[|\tilde{W}_s^T(k)|^2 \overline{S}_s^2 + \delta_s^2 + 2\delta_s w_s \overline{S}_s\right]}{\bar{\beta}}} \right\}.\end{aligned}$$

Notice that δ_s is an arbitrarily small positive constant. Further, selecting β close to unity, then $\bar{\beta} \simeq -1$ and $\mathcal{E}_s$ becomes $\mathcal{O}(|\tilde{W}_s|)$. Unfortunately, choosing $\beta \simeq 1$ has the negative effect of slowing down the adaptation of $\hat{W}_s$ given by (8.9). Obviously, a compromise between the size of $\mathcal{E}_s$ and the adaptability of $\hat{W}_s$ is necessary. We have thus proven the theorem:

Theorem 8.1 *Consider the system (8.2). The group sending time estimation algorithm (8.3), (8.9), (8.11) with $\beta \in \left(1 - \sqrt{1/(1 + \overline{S}_s^2)}, 1\right) \bigcup \left(1, 1 + \sqrt{1/(1 + \overline{S}_s^2)}\right)$, $|S_s(D(k))| \le \overline{S}_s$ guarantees the uniform ultimate boundedness of the group sending time estimation error $e_s(k) = y_s(k) - \hat{y}_s(k)$, with respect to the set $\mathcal{E}_s$, which can be made $\mathcal{O}(w_s)$ by appropriately selecting β.*

Given the estimated, at group k, mean round trip time value $\widehat{RTT}_s(k)$, we take the desired round trip time for all packets within group k to be equal to

$$RTT_d = \widehat{RTT}_s(k). \tag{8.12}$$

Owing to the dynamic character of the Internet, the almost constant $\hat{p}_s$ value assumption during group transmission is highly unrealistic, especially when the number of packets in the group M is significant. To alleviate this problem we have proposed in [130] the constant monitoring of the $\hat{p}_s$ per packet value in the group and whenever a norm of the difference between the current $\hat{p}_s$ value and the one predicted at the beginning of the group transmission exceeds a certain predetermined threshold $d\hat{p}_s$, the group transmission is terminated and the procedure is repeated by sending the next M packets.

Remark 8.1 *Note from (8.12) that for each packet group, the corresponding value of RTT_d is kept constant, thereby ensuring the piecewise nature of the RTT_d, as required by Remark 7.3.*

Remark 8.2 *Theorem 8.1 guarantees the uniform ultimate boundedness of e_s with respect to $\mathcal{E}_s$, thus always providing a bounded reference to the rate controller. However, $\mathcal{E}_s$ cannot be made arbitrarily small without further imposing a persistency of excitation assumption on $\hat{W}_s$, whose satisfaction unfortunately cannot be verified a priori. In the worst case large e_s will guide the rate controller to achieve a slower group sending time than what is feasible. Reducing the size of $\mathcal{E}_s$ deserves further investigation.*

Remark 8.3 *The round trip time per packet modeling was based on a fluid flow model of RT. However, we didn't follow the same approach when developing the group sending time estimation algorithm, where a discrete time model was adopted, since the latter operates on groups of packets, remaining idle during group transmission. Thus, the RTT_d estimator operates at a higher level of hierarchy than the future path congestion level estimator and the source rate controller.*

In Table 8.1 the feasible desired round trip time estimator design is summarized.

8.2.1 Proof of Lemma 8.1

The proof can be readily established by noting that $\hat{W}_s(k + 1) \in \mathcal{R}_s$ when $\hat{W}_s(k) \in \mathcal{R}_s$. Employing the adaptive law (8.11), we obtain the following cases:

Case 1 ($\|\hat{W}_s(k) + (1 - \beta)e_s(k)S_s(D(k))\| \leq w_s$): Then $\|\hat{W}_s(k + 1)\| \leq w_s$

Case 2 ($\|\hat{W}_s(k) + (1 - \beta)e_s(k)S_s(D(k))\| > w_s$ and $(1 - \beta)e_s(k)S_s^T(D(k))$ $\hat{W}_s(k) \geq 0$): Then

$$
\begin{aligned}
\|\hat{W}_s(k + 1)\| &= \left\| \hat{W}_s(k) + (w_s - \|\hat{W}_s(k)\|) \frac{(1 - \beta)e_s(k)S_s(D(k))}{\|(1 - \beta)e_s(k)S_s(D(k))\|} \right\| \\
&\leq \left\| \hat{W}_s(k) + \left(w_s - \|\hat{W}_s(k)\|\right) \frac{\hat{W}_s(k)}{\|\hat{W}_s(k)\|} \right\|
\end{aligned}
$$

since $\forall$ a,b,c $\in \Re^n$, $\|a + b\| \leq \|a + c\|$ whenever $\|b\| = \|c\|$ and a,b are of the same direction. Hence,

$$
\|\hat{W}_s(k + 1)\| \leq \left\| w_s \frac{\hat{W}_s(k)}{\|\hat{W}_s(k)\|} \right\| = w_s.
$$

TABLE 8.1

Feasible desired round trip time estimator

Actual System:	$\overline{RTT}_s(k+1) = \overline{RTT}_s(k) + W_s^{*T}S_s(D(k)) + \omega_s(D(k))$
	$D(k) = \{\overline{RTT}_s(k), \overline{RTT}_s(k-1), ..., \overline{RTT}_s(k-m_1),$
	$\hat{p}_s(k), \hat{p}_s(k-1), \hat{p}_s(k-2), ..., \hat{p}_s(k-m_2)\}$
Error System:	$e_s(k+1) = (1-\beta)e(k) - (\hat{W}_s(k) - W_s^*)^T S_s(D(k))$
	$+\omega_s(D(k))$
	$e_s(k) = \overline{RTT}_s(k) - \hat{\overline{RTT}}_s(k)$
Control Law:	$\hat{\overline{RTT}}_s(k+1) = \hat{\overline{RTT}}_s(k) + \hat{W}_s^T(k)S_s(D(k)) + \beta e_s(k)$
Update Law:	$\hat{W}_s(k+1) = \hat{W}_s(k) + \mathcal{P}_s\{(1-\beta)e_s(k)S_s(D(k))\}$
	$\mathcal{P}_s\{z\} = \begin{cases} z & \text{if } \|\hat{W}_s(k) + z\| \le w_s \\ z - z\left(1 - \frac{w_s - \|\hat{W}_s(k)\|}{\|z\|}\right) & \text{if } \|\hat{W}_s(k) + z\| > w_s \\ & \text{and } z\hat{W}_s(k) \ge 0 \end{cases}$
Properties:	e_s is u.u.b with respect to $\mathcal{E}_s = \{e_s(k):$
	$\|e_s(k)\| < \frac{-\|1-\beta\|(\delta_s + w_s \overline{S}_s)}{\beta}$
	$+\sqrt{\frac{(\delta_s + w_s \overline{S}_s)^2\|1-\beta\|^2}{\beta^2} - \frac{[\|\tilde{W}_s^T(k)\|^2 \overline{S}_s^2 + \delta_s^2 + 2\delta_s w_s \overline{S}_s]}{\beta}}\}$
Conditions:	$W_c \in \Re^4,\ S_s(D(k)) \in \Re^4,\ D(k) = \{\overline{RTT}_s(k), \overline{RTT}_s(k-1),$
	$..., \overline{RTT}_s(k-m_1), \hat{p}_s(k), \hat{p}_s(k-1), ..., \hat{p}_s(k-m_2)\},$
	$m_1 \in \aleph,\ m_2 \in \aleph$
Requirements:	$\beta \in \left(1 - \sqrt{1/(1+\overline{S}_s^2)}, 1\right) \bigcup \left(1, 1 + \sqrt{1/(1+\overline{S}_s^2)}\right),$
	$\|S_s(D(k))\| \le \overline{S}_s$

8.2.2 Proof of Lemma 8.2

Substituting (8.3)-(8.6) and (8.9),(8.11) in (8.8) we finally obtain

$$\Delta J = \beta(\beta - 2)e_s^2(k)(1-\beta)^2 e_s^2(k)|S_s(D(k))|^2$$
$$+ 2(1-\beta)e_s(k)\omega_s(D(k)) + |\tilde{W}_s^T(k)S_s(D(k))|^2 + \omega_s^2$$
$$- 2\omega_s(D(k))\tilde{W}_s^T(k)S_s(D(k)) + \mathcal{I}_{ind}\Delta J_1$$

where $\mathcal{I}_{ind}$ is an indicator function defined as $\mathcal{I}_{ind} = 1$ if the projection operation is active and $\mathcal{I}_{ind} = 0$ otherwise. Further,

$$
\begin{aligned}
\Delta J_1 &= (1-\beta)^2 e_s(k)^2 \|S_s(D(k))\|^2 \left(1 - \frac{w_s - \|\hat{W}_s(k)\|}{\|(1-\beta)e_s(k)S_s(D(k))\|}\right)^2 \\
&\quad -2(1-\beta)^2 e_s(k)^2 \|S_s(D(k))\|^2 \left(1 - \frac{w_s - \|\hat{W}_s(k)\|}{\|(1-\beta)e_s(k)S_s(D(k))\|}\right) \\
&\quad -2(1-\beta)e_s(k)S_s^T(D(k)) \left(1 - \frac{w_s - \|\hat{W}_s(k)\|}{\|(1-\beta)e_s(k)S_s(D(k))\|}\right) \hat{W}_s(k).
\end{aligned}
$$

Moreover,

$$
\begin{aligned}
w_s &< \|\hat{W}_s(k) + (1-\beta)e_s(k)S_s(D(k))\| \\
&\leq \|\hat{W}_s(k)\| + \|(1-\beta)e_s(k)S_s(D(k))\|.
\end{aligned}
$$

Hence

$$
0 < \left(1 - \frac{w_s - \|\hat{W}_s(k)\|}{\|(1-\beta)e_s(k)S_s(D(k))\|}\right) < 1
$$

and

$$
\left(1 - \frac{w_s - \|\hat{W}_s(k)\|}{\|(1-\beta)e_s(k)S_s(D(k))\|}\right)^2 \leq \left(1 - \frac{w_s - \|\hat{W}_s(k)\|}{\|(1-\beta)e_s(k)S_s(D(k))\|}\right).
$$

Therefore,

$$
\begin{aligned}
\Delta J_1 &= -(1-\beta)^2 e_s(k)^2 \|S_s(D(k))\|^2 \left(1 - \frac{w_s - \|\hat{W}_s(k)\|}{\|(1-\beta)e_s(k)S_s(D(k))\|}\right) \\
&\quad -2(1-\beta)e_s(k)S_s^T(D(k)) \left(1 - \frac{w_s - \|\hat{W}_s(k)\|}{\|(1-\beta)e_s(k)S_s(D(k))\|}\right) \hat{W}_s(k).
\end{aligned}
$$

Furthermore, the term

$$
-2\left(1 - \frac{w_s - \|\hat{W}_s(k)\|}{\|(1-\beta)e_s(k)S_s(D(k))\|}\right)(1-\beta)e_s(k)S_s^T(D(k))\hat{W}_s(k)
$$

is negative by definition.

Hence, ΔJ_1 is negative, which implies that ΔJ becomes more negative, whenever projection modification is applied.

8.3   Rate Control Design

In this section, we shall present a systematic tool, based upon a Lyapunov function derivative estimation approach, for the design of controllers capable to guarantee a uniform ultimate boundedness property for the tracking error $e = RTT - RTT_d$, as well as the uniform boundedness of all other signals in the closed-loop.

We recall from Chapter 7 that

$$\dot{RTT} = f(RTT, u, p), \quad RTT \in [0, \tau_0], \quad u \in (0, \overline{u}], \quad p \in [0, 1). \tag{8.13}$$

Owing to the smoothness of $f(\cdot)$ there exist smooth functions f_1, f_2 such that

$$\dot{RTT} = f_1(RTT, u)f_2(p) \tag{8.14}$$

$\forall \, RTT \in [0, \tau_0], \, \forall \, u \in (0, \overline{u}], \, \forall \, p \in [0, 1).$

Remark 8.4 *Notice that (8.14) may still preserve all qualitative RTT per packet characteristics such as:*

- *$RTT \rightarrow \tau_0$ when $L(S, D)$ is congested $(p = 1)$ making (8.14) practically uncontrollable.*

- *Enlargement of p $(p \in [0, 1))$ should progressively lead to attenuation of RTT dependency on u.*

Hence, generality is not harmed, which is also verified in the simulation studies performed at the end of the chapter.

Employing (7.8), $f_2(p)$ becomes

$$f_2(\hat{p}_s + \tilde{p}) = f_{21}(\hat{p}_s) + f_{22}(\hat{p}_s, \tilde{p}) \tag{8.15}$$

where $f_{21}(\hat{p}_s)$, $f_{22}(\hat{p}_s, \tilde{p})$ are unknown, smooth and bounded functions, with f_{22} having the property $\lim_{\tilde{p} \to 0} f_{22}(\hat{p}_s, \tilde{p}) = 0$. In the special case where $\tilde{p} = 0$, $f_2(p) \equiv f_2(\hat{p}_s)$. Substituting (8.15) into (8.14) we obtain:

$$\dot{RTT} = f_1(RTT, u)f_{21}(\hat{p}_s) + f_1(RTT, u)f_{22}(\hat{p}_s, \tilde{p}). \tag{8.16}$$

Notice that $\tilde{p}$ is not available for measurement and thus cannot be used for control. The following assumptions are necessary in the subsequent analysis.

Assumption 8.1 *There exist some unknown $\mathcal{K}$-functions g_i, $i = 1, 2, 3$ such that*

$$f_1(RTT, u) \leq g_1(RTT, u), \qquad \forall RTT \in [0, \tau_0], \quad \forall u \in [0, \overline{u}]$$
$$f_{22}(\hat{p}_s, \tilde{p}) \leq g_2(\hat{p}_s)g_3(\tilde{p}), \qquad \forall p \in [0, 1).$$

Moreover, since g_3 is a $\mathcal{K}$-function there exists an unknown positive constant d such that

$$g_3(\tilde{p}) \le d. \tag{8.17}$$

Remark 8.5 *Notice that decompositions (8.14), (8.16) as well as Assumption 1 do not harm generality owing to the unknown, smooth and bounded character of f_1, f_2, f_{21}, f_{22}, g_i, $i = 1, 2, 3$ and the fact that RTT, u, p, $\hat{p}_s$ belong to the bounded sets $[0, \tau_0]$, $(0, \overline{u}]$, $[0, 1]$, $[0, 1]$, respectively.*

Assumption 8.2 *The solution of (8.16) can be forced to be uniformly ultimately bounded with respect to an arbitrarily small neighborhood of $e = 0$.*

Owing to Assumption 8.2, there exists a radially unbounded robust control Lyapunov function $V(e) : \Re \to \Re_+$ satisfying $\lambda_1 |e|^2 \le V(e) \le \lambda_1 |e|^2$, $\lambda_1, \lambda_2 > 0$ and a control input u such that:

$$\dot{V} = \frac{\partial V(e)}{\partial e} \left[f_1(RTT, u) f_{21}(\hat{p}_s) + f_1(RTT, u) f_{22}(\hat{p}_s, \tilde{p}) \right] \le 0 \tag{8.18}$$

$\forall RTT, u, p, \hat{p}_s \in \mathcal{A}$, where the set $\mathcal{A}$ is defined as $\mathcal{A} = \{RTT, u, p, \hat{p}_s \in \Re : 0 \le RTT \le \tau_0, \ u_0 \le u \le \overline{u}, \ 0 \le p < 1, \ 0 \le \hat{p}_s < 1\}$ with u_0 an arbitrarily small positive constant.

To proceed, let us define:

$$A(e, RTT, u, \hat{p}_s) = \frac{\partial V(e)}{\partial e} f_1(RTT, u) f_{21}(\hat{p}_s) \tag{8.19}$$

$$B(e, RTT, u, \hat{p}_s) = \left| \frac{\partial V(e)}{\partial e} \right| g_1(RTT, u) g_2(\hat{p}_s) d. \tag{8.20}$$

Since $f_1(RTT, u)$, $f_{21}(\hat{p}_s)$, $g_1(RTT, u)$, $g_2(\hat{p}_s)$, d and $V(e)$ are assumed unknown, we may utilize the neural nets density property and substitute the highly uncertain terms A, B in (8.19), (B) by linear in the weights neural networks, plus a modeling error term $\forall e, RTT, u, \hat{p}_s \in \Omega \subset \Xi$ where $\Xi = \{e, RTT, u, \hat{p}_s \in \Re : \ 0 \le RTT \le \tau_0, \ u_0 \le u \le \overline{u}, \ 0 \le \hat{p}_s < 1\}$. In other terms, there exist constant but unknown weight values W_1^*, W_2^* such that

$$A(e, RTT, u, \hat{p}_s) = W_1^{*T} S_1(e, RTT, u, \hat{p}_s) + \omega_1(e, RTT, u, \hat{p}_s) \tag{8.21}$$

$$B(e, RTT, u, \hat{p}_s) = W_2^{*T} S_2(e, RTT, u, \hat{p}_s) + \omega_2(e, RTT, u, \hat{p}_s). \tag{8.22}$$

Such a substitution is possible owing to the smoothness and boundness of $A(e, RTT, u, \hat{p}_s)$, $B(e, RTT, u, \hat{p}_s)$ for all $e, RTT, u, \hat{p}_s \in \Omega \subset \Xi$. The following assumption is reasonable owing to the approximation capabilities of the linear in the weights neural nets.

Assumption 8.3 *In a compact region $\Omega \subset \Re^4$, $|\omega_1(e, RTT, u, \hat{p}_s)| \le \delta_1$, $|\omega_2(e, RTT, u, \hat{p}_s)| \le \delta_2$ where $\delta_1, \delta_2 \ge 0$ are unknown but small bounds.*

Furthermore, the results obtained in this paper are semiglobal in the sense that they are valid as long as e, RTT, u, $\hat{p}_s$ remains in Ω, where the set Ω can be made arbitrarily large. If Assumption 8.3 holds for all e, RTT, u, $\hat{p}_s \in \Xi$, then the results become global.

Substituting (8.19) and (B) in (9.9) and employing (8.21) and (8.22), $\dot{V}$ finally becomes

$$
\begin{aligned}
\dot{V} \leq\ & W_1^{*T} S_1(e, RTT, u, \hat{p}_s) + W_2^{*T} S_2(e, RTT, u, \hat{p}_s) \\
& + \omega_1(e, RTT, u, \hat{p}_s) \\
& + \omega_2(e, RTT, u, \hat{p}_s), \forall RTT, u, p, \hat{p}_s \in \mathcal{A}.
\end{aligned} \tag{8.23}
$$

Let us now consider the Lyapunov function candidate:

$$
L = kV(e) + \frac{1}{2}|\tilde{W}_1|^2 + \frac{1}{2}|\tilde{W}_2|^2 + \frac{1}{2}u^2
$$

where $V(e)$ is the previously defined robust control Lyapunov function. The parameter errors $\tilde{W}_i$, $i = 1, 2$ are defined as $\tilde{W}_i = W - W^*$, $i = 1, 2$.

Differentiating with respect to time along the solutions of (8.16) we obtain

$$
\begin{aligned}
\dot{L} =\ & k\dot{V}(e) + \tilde{W}_1^T \dot{W}_1 + \tilde{W}_2^T \dot{W}_2 + u\dot{u} \\
=\ & k\frac{\partial V(e)}{\partial e} f_1(RTT, u) f_{21}(\hat{p}_s) \\
& + k\frac{\partial V(e)}{\partial e} f_1(RTT, u) f_{22}(\hat{p}_s, \tilde{p}) \\
& + \tilde{W}_1^T \dot{W}_1 + \tilde{W}_2^T \dot{W}_2 + u\dot{u}.
\end{aligned}
$$

Using Assumption 8.1 and (8.17) we obtain:

$$
\begin{aligned}
\dot{L} \leq\ & k\frac{\partial V(e)}{\partial e} f_1(RTT, u) f_{21}(\hat{p}_s) \\
& + k\left|\frac{\partial V(e)}{\partial e}\right| g_1(RTT, u) g_2(\hat{p}_s) d \\
& + \tilde{W}_1^T \dot{W}_1 + \tilde{W}_2^T \dot{W}_2 + u\dot{u}.
\end{aligned}
$$

Further, employing (8.23) we get

$$
\begin{aligned}
\dot{L} \leq\ & kW_1^{*T} S_1(e, RTT, u, \hat{p}_s) + kW_2^{*T} S_2(e, RTT, u, \hat{p}_s) \\
& + k\omega(e, RTT, u, \hat{p}_s) + \tilde{W}_1^T \dot{W}_1 + \tilde{W}_2^T \dot{W}_2 + u\dot{u}.
\end{aligned}
$$

Owing to Assumption 3 we conclude that the generalized modeling error

$$
\omega(e, RTT, u, \hat{p}_s) = \omega_1(e, RTT, u, \hat{p}_s) + \omega_2(e, RTT, u, \hat{p}_s)
$$

is uniformly bounded by a small yet unknown constant $\forall\, e, RTT, u, \hat{p}_s \in \Omega$. Specifically:

$$
|\omega(e, RTT, u, \hat{p}_s)| \leq |\omega_1(e, RTT, u, \hat{p}_s)| + |\omega_2(e, RTT, u, \hat{p}_s)| = \delta_1 + \delta_2 = \delta.
$$

Choosing:

$$\dot{u} = \frac{1}{u}\left[-kW_1^T S_1(e, RTT, u, \hat{p}_s) - kW_2^T S_2(e, RTT, u, \hat{p}_s)\right.$$
$$\left. -\gamma_1 e^2 - \gamma_2 u^2\right] \tag{8.24}$$

where $\gamma_1, \gamma_2 > 0$, $\dot{L}$ becomes:

$$\dot{L} \leq -k\tilde{W}_1^{*T} S_1(e, RTT, u, \hat{p}_s) + k\tilde{W}_2^{*T} S_2(e, RTT, u, \hat{p}_s)$$
$$+k\omega(e, RTT, u, \hat{p}_s) - \gamma_1 e^2 - \gamma_2 u^2$$
$$+\tilde{W}_1^T \dot{W}_1 + \tilde{W}_2^T \dot{W}_2.$$

Moreover, if:

$$\dot{W}_i = -k_i W_i + k S_i(e, RTT, u, \hat{p}_s), \quad k_i > 0, \ i = 1, 2 \tag{8.25}$$

then

$$\dot{L} \leq k\omega(e, RTT, u, \hat{p}_s) - \gamma_1 e^2 - \gamma_2 u^2 - k_1 \tilde{W}_1^T W_1 - k_2 \tilde{W}_2^T W_2. \tag{8.26}$$

Further, we recall that for $z, \tilde{z} \in \Re^n$, with $\tilde{z} = z - z^*$ it is true that

$$\tilde{z}^T z = \frac{1}{2}|\tilde{z}|^2 + \frac{1}{2}|z|^2 - \frac{1}{2}|z^*|^2.$$

Hence,

$$\dot{L} \leq k\delta - \gamma_1 e^2 - \gamma_2 u^2 - \sum_{i=1}^{2}\left[\frac{k_i}{2}\left|\tilde{W}_i\right|^2 + \frac{k_i}{2}|W_i|^2 - \frac{k_i}{2}|W_i^*|^2\right]$$

$$\leq -\gamma_1 e^2 - \gamma_2 u^2 - \sum_{i=1}^{2}\left[\frac{k_i}{2}\left|\tilde{W}_i\right|^2\right] + \mu$$

where $\mu = k\delta + \sum_{i=1}^{2}\left[\frac{k_i}{2}|W_i^*|^2\right]$. From the definition of $V(e)$ we argue that there exist values of γ_1 such that $\gamma_1|e|^2 \geq kV(e), \forall e \in [-\tau_0, \tau_0]$. Define $c = \min\{1, k_1, k_2, \gamma_2\}$. Then $\dot{L}$ becomes:

$$\dot{L} = -cL + \mu$$

which yields:

$$0 \leq L(t) \leq \frac{\mu}{c} + \left[L(0) - \frac{\mu}{c}\right]e^{-ct}. \tag{8.27}$$

Hence, we conclude the uniform boundedness of RTT, u, W_1, W_2. Moreover, after dropping the negative terms $-\gamma_2 u^2$, $-\frac{k_i}{2}\left|\tilde{W}_i\right|$, $i = 1, 2$ we obtain

$$\dot{L} \leq -\gamma_1 e^2 + \mu \leq 0 \tag{8.28}$$

provided that

$$|e| > \sqrt{\mu/\gamma_1}.$$

Thus, the *RTT* per packet error possesses a uniform ultimate boundedness property with respect to the set

$$\mathcal{E} = \left\{ e(k) \in \Re : |e| < \sqrt{\mu/\gamma_1} \right\} \tag{8.29}$$

whose size may be arbitrarily small either by enlarging γ_1, or by reducing the magnitude of k, k_1, k_2. The aforementioned analysis is summarized in the theorem that follows.

Theorem 8.2 *Consider the round trip time system (8.16) which satisfies Assumptions 8.1-8.3. The rate controller (8.24) together with the update laws (8.25) guarantee the uniform ultimate boundedness of the packet round trip error with respect to the arbitrarily small set (8.29), as well as the uniform boundedness of all other signals in the closed loop.*

The regulation and stability results proven in Theorem 8.2 are valid provided $e, RTT, u, \hat{p}_s \in \Omega \subset \Xi$ throughout transmission. Notice that $\hat{p}_s \in [0, 1)$ by construction. To verify that $e \in \Omega_e \subset [-\tau_0, \tau_0]$ and thus $RTT \in \Omega_{RTT} \subset [0, \tau_0]$, we observe that there exist design constants $\gamma, k, k_1, k_2 > 0$ to guarantee that $|e| \leq \tau_0$. Hence, we argue that if we start with an initial condition $e(0) \in \mathcal{E}$, then owing to the uniform ultimate boundedness of e we have that $e \in \mathcal{E} \subset \Omega_e \subset [-\tau_0, \tau_0]$ for every transmitted packet. If however $e(0) \in \Omega_e/\mathcal{E}$ we distinguish two cases:

Case 1 $(L(0) > \frac{\mu}{c})$: From (8.27) and the definition of L, we conclude that:

$$kV(e) \leq L \leq L(0)$$

or

$$|e| \leq V^1 \left(V(0) + \frac{1}{2k} \left| \tilde{W}_1(0) \right|^2 + \frac{1}{2k} \left| \tilde{W}_2(0) \right|^2 + \frac{1}{2k} u^2(0) \right).$$

Thus there exist a k for which

$$\left\{ e \in \Re : |e| \leq V^{-1} \left(V(0) + \frac{1}{2k} \left| \tilde{W}_1(0) \right|^2 + \frac{1}{2k} \left| \tilde{W}_2(0) \right|^2 + \frac{1}{2k} u^2(0) \right) \right\} \subset \Omega_e.$$

Case 2 $(L(0) \leq \frac{\mu}{c})$: In this case we obtain:

$$kV(e) \leq L \leq \frac{\mu}{c}$$

or

$$|e| \leq V^{-1} \left(\frac{\mu}{kc} \right).$$

Hence, similar to Case 1, there exist values of $k, k_1, k_2 > 0$ that makes:

$$\left\{ e \in \Re : \ |e| \leq V^{-1}\left(\frac{\mu}{kc}\right) \right\} \subset \Omega_e.$$

Thus, in any case we can always find design constants to guarantee that $e \in \Omega_e \subset [-\tau_0, \tau_0]$ for every transmitted packet.

8.3.1 Guaranteeing Boundness of Transmission Rate

Previously we have shown the uniform boundedness of u. However, since the proposed rate controller (8.24), (8.25), (8.26) is dynamic, there is no guarantee that its output will be strictly confined in $[u_0, \ \overline{u}]$. In this subsection we present stable modifications that alleviate this issue.

We assume that we start with an initial condition $u(0) \in [u_0, \ \overline{u}]$ and we distinguish two cases.

Case 1 $(u = \overline{u})$: The rate controller has reached its upper limit. Now (8.24) is modified to:

$$\dot{u} = \frac{1}{u}\left[-kW_1^T S_1(e, RTT, u, \hat{p}_s) - kW_2^T S_2(e, RTT, u, \hat{p}_s) \right. $$
$$\left. -\gamma_1 e^2 - \gamma_2 u^2 + \psi\right] \tag{8.30}$$

where

$$\psi = \begin{cases} -M, & \text{if } u = \overline{u} \text{ and } M - \gamma_1 e^2 - \gamma_2 u^2 > 0 \\ \\ 0, & \text{otherwise} \end{cases} \tag{8.31}$$

with $M = -kW_1^T S_1(e, RTT, u, \hat{p}_s) - kW_2^T S_2(e, RTT, u, \hat{p}_s)$. Employing (8.30), (8.31) it is straightforward to conclude that whenever u has reached its upper limit and has a tendency to move upwards (i.e., $u = \overline{u}$ and $\dot{u} > 0$), then $\dot{u}$ becomes:

$$\dot{u} = \frac{1}{u}\left[-\gamma_1 e^2 - \gamma_2 u^2\right] < 0$$

guaranteeing $u \leq \overline{u}$, for all transmitted packets. Furthermore, (8.30), (8.31) do not alter the stability properties proven thus far, since $\dot{L}$ is augmented with the extra term:

$$\psi = kW_1^T S_1(e, RTT, u, \hat{p}_s) + kW_2^T S_2(e, RTT, u, \hat{p}_s) < -\gamma_1 e^2 - \gamma_2 u^2 < 0.$$

Case 2 $(u = u_0)$: The rate controller has reached its lower limit. At this point and whenever there is a tendency to move downwards we substitute the terms $S_i(e, RTT, u, \hat{p}_s)$, $i = 1, 2$ by:

$$S_i(e, RTT, u, \hat{p}_s) = S_{ii}(e, RTT, u, \hat{p}_s)b_i(W_i, e, RTT, u, \hat{p}_s)$$
$$+b_{oi}(W_i, e, RTT, u, \hat{p}_s) \tag{8.32}$$

where

$$
b_i(W_i, e, RTT, u, \hat{p}_s) = \begin{cases} -\dfrac{3(\gamma_1 e^2 + \gamma_2 u^2) W_i^T S_{ii}(e, RTT, u, \hat{p}_s)}{4k \| W_i^T S_{ii}(e, RTT, u, \hat{p}_s) \|^2}, & \text{if } C_1 \\[1em] 1, & \text{otherwise} \end{cases}
$$

$$(8.33)$$

$$
b_{oi}(W_i, e, RTT, u, \hat{p}_s) = \begin{cases} -\dfrac{3(\gamma_1 e^2 + \gamma_2 u^2) W_i}{4k \| W_i \|^2}, & \text{if } C_2 \\[1em] 0, & \text{otherwise} \end{cases} \qquad (8.34)
$$

with

$$
C_1 = \left[u = u_0, \; M - \gamma_1 e^2 - \gamma_2 u^2 < 0 \text{ and } W_i^T S_{ii}(e, RTT, u, \hat{p}_s) \neq 0 \right]
$$

$$
C_2 = \left[u = u_0, \; M - \gamma_1 e^2 - \gamma_2 u^2 < 0 \text{ and } W_i^T S_{ii}(e, RTT, u, \hat{p}_s) = 0 \right]
$$

for $i = 1, 2$.

Remark 8.6 *Notice that whenever $u > u_0$ or if $u = u_0$ and there is a tendency to move upwards $S_i(e, RTT, u, \hat{p}_s)$, $i = 1, 2$ remains unaltered.*

Employing (8.33), (8.34), (8.30) we may straightforwardly conclude that whenever $u = u_0$ and $M - \gamma_1 e^2 - \gamma_2 u^2 < 0$ then:

$$
\dot{u} = \frac{\gamma_1 e^2 + \gamma_2 u^2}{2u} > 0.
$$

However, (8.34) is valid only when $\| W_i \| \neq 0$, $i = 1, 2$ which can be satisfied by applying the standard projection modification [137] on W_i, $i = 1, 2$.

Hence, in both cases $\dot{u}$ is modified to guarantee that $u \in [u_0, \overline{u}]$ for every transmitted packet.

8.3.2   Reducing Rate in Congestion

Whenever the source detects congestion in its path ($\hat{p}_s = 1$), the only acceptable policy would be to reduce its rate until $\hat{p}_s < 1$.

To guarantee that $\dot{u} \leq 0$ whenever $\hat{p}_s = 1$ (with $\dot{u} = 0$ only when the rate controller has reached its lower limit $u = u_0$), we modify (8.30) as follows:

$$
\begin{aligned}
\dot{u} = \; & \frac{1}{u} \Big[-kW_1^T S_1(e, RTT, u, \hat{p}_s) - kW_2^T S_2(e, RTT, u, \hat{p}_s) \\
& - \gamma_1 e^2 - \gamma_2 u^2 + \psi + \varphi \Big]
\end{aligned}
$$

$$(8.35)$$

where

$$
\varphi = \begin{cases} -M, & \text{if } \hat{p}_s = 1,\ u_0 < u < \bar{u} \text{ and } M_1 > 0 \\[2mm] -\frac{1}{2}(\gamma_1 e^2 + \gamma_2 u^2), & \text{if } \hat{p}_s = 1,\ u = u_0 \text{ and } M_1 < 0 \\[2mm] -M + \gamma_1 e^2 + \gamma_2 u^2, & \text{if } \hat{p}_s = 1,\ u = u_0 \text{ and } M_1 > 0 \\[2mm] 0, & \text{otherwise} \end{cases} \tag{8.36}
$$

with $M_1 = \left[M - \gamma_1 e^2 - \gamma_2 u^2 \right]$.

To understand the operation of the proposed modification, we distinguish three cases.

Case 1 ($u_0 < u < \bar{u}$ and $M - \gamma_1 e^2 - \gamma_2 u^2 > 0$): In this case, $u \in (u_0,\ \bar{u})$ when $\hat{p}_s = 1$ and at the same time u has a tendency to move upwards. Notice that $\psi = 0$ in this case. Employing (8.36) into (8.35), we obtain:

$$
\dot{u} = -\frac{\gamma_1 e^2 + \gamma_2 u^2}{u} < 0.
$$

Case 2 ($u = u_0$ and $M - \gamma_1 e^2 - \gamma_2 u^2 < 0$): Employing (8.33), (8.34), (8.36), (8.35) we may straightforwardly conclude that:

$$
\dot{u} = 0
$$

since $\psi = 0$ for this case also.

Case 3 ($u = u_0$ and $M - \gamma_1 e^2 - \gamma_2 u^2 > 0$): Employing (8.36), (8.35) we may straightforwardly conclude that:

$$
\dot{u} = 0.
$$

Furthermore, since in each case, we add the negative terms $-M$, $-\frac{1}{2}(\gamma_1 e^2 + \gamma_2 u^2)$ and $-M + \gamma_1 e^2 + \gamma_2 u^2$, respectively, in $\dot{L}$, the stability properties proven thus far are not harmed.

The results of the rate controller design are summarized in Table 8.2.

TABLE 8.2
Rate controller

Actual System:	$\dot{RTT} = f(RTT, u, p), \quad RTT \in [0, \tau_0], \quad u \in (0, \bar{u}], \quad p \in [0, 1)$

Control Law:

$$\dot{u} = \frac{1}{u}\left[-kW_1^T S_1(e, RTT, u, \hat{p}_s) - kW_2^T S_2(e, RTT, u, \hat{p}_s)\right.$$
$$\left. -\gamma_1 e^2 - \gamma_2 u^2 + \psi + \varphi\right]$$

$$\psi = \begin{cases} -M, & \text{if } u = \bar{u} \text{ and } M - \gamma_1 e^2 - \gamma_2 u^2 > 0 \\ 0, & \text{otherwise} \end{cases}$$

$$\varphi = \begin{cases} -M, & \text{if } \hat{p}_s = 1, u_0 < u < \bar{u} \text{ and } M_1 > 0 \\ -\frac{1}{2}(\bar{\gamma}), & \text{if } \hat{p}_s = 1, u = u_0 \text{ and } M_1 < 0 \\ -M + \bar{\gamma}, & \text{if } \hat{p}_s = 1, u = u_0 \text{ and } M_1 > 0 \\ 0, & \text{otherwise} \end{cases}$$

$$M = -kW_1^T S_1(e, RTT, u, \hat{p}_s) - kW_2^T S_2(e, RTT, u, \hat{p}_s)$$
$$M_1 = \left[M - \gamma_1 e^2 - \gamma_2 u^2\right]$$
$$\bar{\gamma} = \gamma_1 e^2 + \gamma_2 u^2$$

Update Laws:

$$\dot{W}_i = -k_i W_i + kS_i(e, RTT, u, \hat{p}_s), \quad k_i > 0, \quad i = 1, 2$$
$$S_i(e, RTT, u, \hat{p}_s) = S_{ii}(e, RTT, u, \hat{p}_s)b_i(W_i, e, RTT, u, \hat{p}_s)$$
$$+ b_{oi}(W_i, e, RTT, u, \hat{p}_s)$$

$$b_i(B) = \begin{cases} -\frac{3(\gamma_1 e^2 + \gamma_2 u^2)W_i^T S_{ii}(e, RTT, u, \hat{p}_s)}{4k\|W_i^T S_{ii}(e, RTT, u, \hat{p}_s)\|^2}, & \text{if } C_1 \\ 1, & \text{otherwise} \end{cases}$$

$$b_{oi}(B) = \begin{cases} -\frac{3(\gamma_1 e^2 + \gamma_2 u^2)W_i}{4k\|W_i\|^2}, & \text{if } C_2 \\ 0, & \text{otherwise} \end{cases}$$

$$B = (W_i, e, RTT, u, \hat{p}_s)$$
$$C_1 = \left[u = u_0, M_1 < 0 \text{ and } W_i^T S_{ii}(e, RTT, u, \hat{p}_s) \neq 0\right]$$
$$C_2 = \left[u = u_0, M_1 < 0 \text{ and } W_i^T S_{ii}(e, RTT, u, \hat{p}_s) = 0\right]$$
for $i = 1, 2$

Properties:

e is u.u.b. with respect to
$$\mathcal{E} = \left\{e(k) \in \Re : |e| < \sqrt{\mu/\gamma_1}\right\}, u_0 \leq u \leq \bar{u}$$

Requirements:

$k_i > 0, \gamma_i > 0, i = 1, 2.$

8.4 Illustrative Example

To evaluate the proposed rate controller, the design methodology developed herein is illustrated via simulations performed on a simple network configuration. To exhibit the performance of the presented scheme two simulation scenarios have been constructed. In the first, the behavior of the rate controller is verified in normal conditions, while in the second its robustness is checked in the presence of strong congestion.

8.4.1 Implementation Details

The desired round trip time estimator and the rate controller designed earlier in this chapter are implemented with:

$$
S_s(D(k)) = \begin{bmatrix} S_1(\overline{RTT}_s(k))S_4(\hat{p}_s(k)) & S_2(\overline{RTT}_s(k-1))S_4(\hat{p}_s(k)) \\ S_3(\overline{RTT}_s(k-2))S_4(\hat{p}_s(k)) & S_4(\hat{p}_s(k)) \end{bmatrix}
$$

$$
S_i(\overline{RTT}_s(k-i)) = \left(\frac{1}{1+e^{-40(\overline{RTT}_s(k+1-i)-0.15)}} \right), \quad i = 0, 1, 2
$$

$$
S_4(\hat{p}_s(k)) = \left(\frac{1}{1+e^{-15(\hat{p}_s(k)-0.5)}} \right)
$$

$$
S_{11}(x,\hat{p}_s) = 0.7 S_{g1}(\hat{p}_s) S_{g2}(x)
$$

$$
S_{g1}(\hat{p}_s) = \left(\frac{1}{1+e^{-11(\hat{p}_s-0.45)}} \right)
$$

$$
S_{g2}(x) = \left[\left(\frac{1}{1+e^{-(x-0.1)}} \right)\left(\frac{1}{1+e^{-(x-0.3)}} \right) \quad \left(\frac{1}{1+e^{-(x-0.2)}} \right)^2 \right.
$$
$$
\left(\frac{1}{1+e^{-(x-0.3)}} \right)^2 \quad \left(\frac{1}{1+e^{-(x-0.4)}} \right)^2 \quad \left(\frac{1}{1+e^{-(x-0.5)}} \right)
$$
$$
\left(\frac{1}{1+e^{-(x-0.6)}} \right)^2 \quad \left(\frac{1}{1+e^{-(x-0.7)}} \right) \quad \left(\frac{1}{1+e^{-(x-0.8)}} \right)^2
$$
$$
\left. \left(\frac{1}{1+e^{-(x-0.9)}} \right)^2 \right]^T
$$

$$
S_{22}(x,\hat{p}_s) = \left(\frac{2}{1+e^{-\hat{p}_s}} \right) - 1.
$$

The design parameters that appear in (8.9), (8.10), (8.24), (8.25) were chosen according to Table 8.3.

TABLE 8.3
Value parameters

β	w_s	γ_1	γ_2	k	k_1	k_2
0.95	0.5	105	45	500	0.0001	0.0001

The initial condition on (8.24) was set equal to 2 Mbps, while the neural network weights were initialized as

$$
W_1(0) = \begin{bmatrix} -0.7 & -0.7 & -0.7 & -0.7 & -0.7 & -0.7 & -0.7 & -0.7 & -0.7 \end{bmatrix}
$$
$$
W_2(0) = 0.4
$$
$$
\hat{W}_s(0) = \begin{bmatrix} 0.1 & 0.1 & 0.1 & 0.1 \end{bmatrix}.
$$

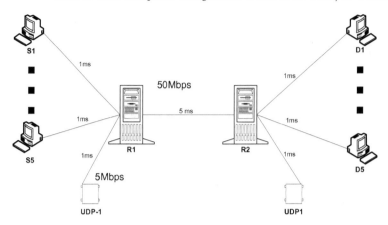

FIGURE 8.1

The single-node network topology.

8.4.2 Network Topology

Consider a simple network topology as shown in Figure 8.1. We have two routers ($R1$ and $R2$) connected via a 50 Mbps, 5 ms link, five *NNRC* sources connected to the routers through independent links and one random constant bit rate (CBR) source. The random CBR source is persistently submitting UDP packets with an average rate of 5 Mbps, having 10% variation, thus constituting the network background traffic. Each pair of source - destination ($Si-Di, i = 1, 2, 3, 4, 5$) connected to the network via a 50 *Mbps* and 1 ms link have one way propagation delay of 7 ms. Each *NNRC* (Si, $i = 1, ..., 5$) source has to transmit 30,000 packets to the corresponding destination (Di, $i = 1, ..., 5$). All packages (*NNRC* and UDP) have a fixed size 12,000 bits. The number of communication channels for each source is constant and set equal to 20 ($n = 20$).

In addition, every link has a buffer with maximum capacity (B_{max}) of 100 packets. All links mark incoming packets with probability p_m equal to

$$p_m = \begin{cases} \dfrac{\text{queue length}}{0.7B_{max}} & \text{, if queue length} < 0.7B_{max} \\ 1 & \text{, otherwise} \end{cases} \qquad (8.37)$$

8.4.3 Normal Scenario

To begin with, we present a normal situation, where all sources in the presence of no external interference succeed the transmission of their packets to the appropriate destinations. The sending times achieved are summarized in Table 8.4. The fair operation of *NNRC* is evident.

TABLE 8.4
Achieved sending times in a normal situation

S_1	S_2	S_3	S_4	S_5
33.403	33.402	33.402	33.403	33.402

During the simulation, the bottleneck utilization (the basic connection between two routers) always stays at 100%, as may be seen from Figure 8.2(a). Bottleneck queue length variations with respect to its maximum capacity are presented in Figure 8.2(b). Notice that during the transition period strong fluctuations are witnessed owing to the simultaneous admittance of all sources in the network. No packet loss was experienced, during transmission. Figure 8.2(c) presents the total transmission rate of the first source. Notice that the achieved throughput is approximately 9 Mbps, which is almost equal to the theoretically expected value obtained from the fair distribution of the available 45 Mbps over five sources. The near perfect allocation of the network bandwidth is dictated by the sending times achieved from the sources.

In Figure 8.3(a) the achieved RTT per packet is pictured for the first channel of the first source, while the corresponding round trip per packet error $e = RT - RTT_d$ is presented in Figure 8.3(b). The performance of the group sending time estimation module, which is responsible for downloading reliable RTT_d values to the rate controller, is demonstrated in Figure 8.3(a). The transmission rate required to achieve such a behavior, as well as the estimated future path congestion level $\hat{p}_s$, are shown in Figures 8.4(b),(a), respectively.

Apparently, after an initial transient period, all errors reach a small neighborhood of zero from which they never escape, as was predicted by the theoretical analysis. Abrupt changes in the rate of transmission of the first channel during the transition period due to strong fluctuations in the level of the queue have the effect of sudden changes in the RTT and e. Furthermore, owing to the piecewise constant nature of RTT_d, the small round trip per packet errors achieved lead to almost constant delays (see Figure 8.3(a)).

Notice that the aforementioned results are indicative. Similar behavior can be shown to hold for the rest of the channels - sources in the network. The almost perfect fair behavior is dictated by the achieved sending times presented in Table 8.4.

8.4.4 Congestion Avoidance Scenario

To better illustrate the operation of the proposed control mechanism in case of congestion, we performed a simulation scenario where the network is driven in a state of strong congestion. The latter is introduced by increasing the CBR source rate from 5 Mbps to 25 Mbps. It is further assumed that congestion appears at t=15s with 1s duration.

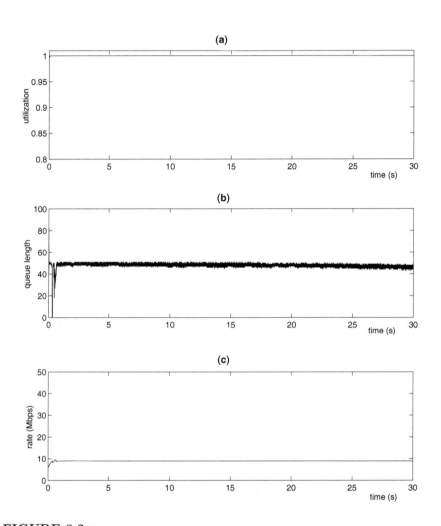

FIGURE 8.2
Performance of *NNRC* in the case of normal scenario: a) utilization of the link
$R1$–$R2$, b) queue length variations of the link $R1$–$R2$, and c) transmission rate
of the first source $S1$.

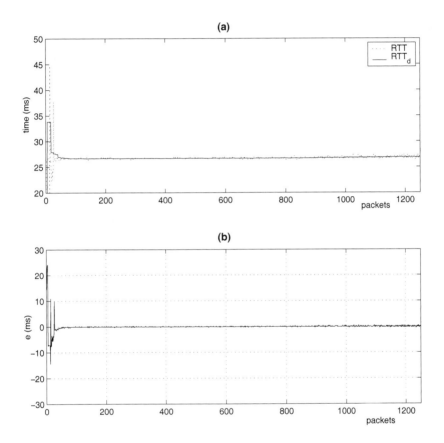

FIGURE 8.3
First channel of the first source in case of normal scenario: a) variations of RTT and RTT_d and b) variations of $e = RTT - RTT_d$.

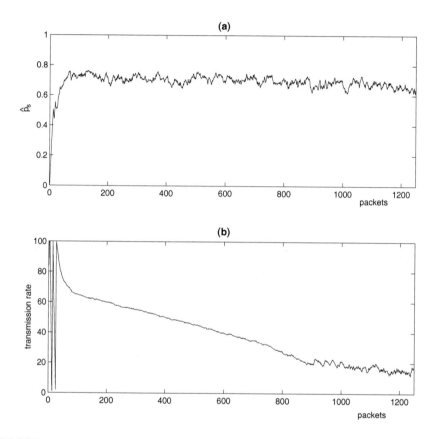

FIGURE 8.4

First channel of the first source in case of normal scenario: a) $\hat{p}_s$ variation and b) transmission rate u.

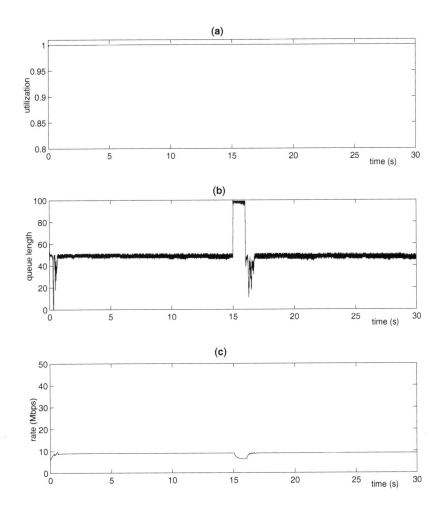

FIGURE 8.5
Performance of *NNRC* in the case of congestion avoidance scenario: a) utilization of the link $R1$–$R2$, b) queue length variations of the link $R1$–$R2$, and c) transmission rate of the first source $S1$.

The sending times achieved are presented in Table 8.5. Besides their expected (owing to the bursty traffic) elevation, compared to the normal scenario, *NNRC* maintains its fair operation.

TABLE 8.5
Achieved sending times congestion avoidance scenario

S_1	S_2	S_3	S_4	S_5
33.770	33.770	33.764	33.769	33,770

During simulation, bottleneck utilization stayed at 100% as is apparent from Figure 8.5(a). Furthermore, the total packet loss was 1396. Figure 8.5(b) presents bottleneck queue length variations with respect to its maximum capacity, while the transmission rate of the first source is illustrated in Figure 8.2(c). Notice the rate reduction when queue length reaches its maximum capacity; a result consistent with the theoretical study.

Figure 8.6(a) shows the RTT per package for the first channel of the first source as well as its corresponding RTT_d. The error $e = RTT - RTT_d$ is shown in Figure 8.6 (b). The transmission rate required to achieve such behavior, and the assessment of path congestion $\hat{p}_s$ for the first source are depicted in Figure 8.7.

Channel rate reduction whenever $\hat{p}_s = 1$ is evident, as was foreseen by the theoretical analysis.

8.5　Concluding Comments

In this chapter attention was paid to the rigorous presentation of the design and operational characteristics of the first two modules of the *NNRC* framework and specifically of a) the feasible round trip time estimator and b) the rate controller. The analysis was carried out under the assumption of a constant and a priori known number of channels.

Robustness against modeling imperfections, exogenous disturbances (UDP traffic) and delays was also performed. The theoretical achievements were further clarified through simulations on a small-scale (though illustrative) example network configuration.

Parts of the results presented in this chapter originally appeared in [130], [131]. To overcome the vast number of uncertainties in the system, the approximation capabilities of linear in the weights neural networks have been employed. However, the whole design is not strongly attached to neural networks. In fact any linear in the parameters function approximation structure (e.g., fuzzy systems, wavelets, polynomials etc.) could be equally used as well.

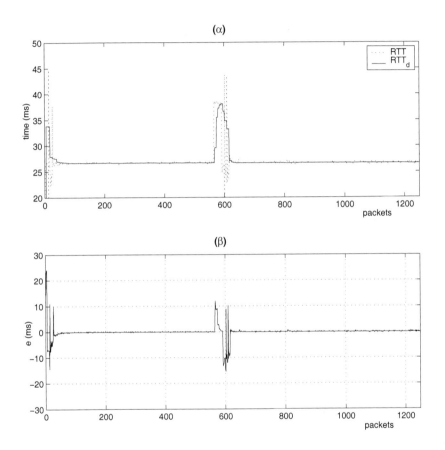

FIGURE 8.6

Picture of the first channel first source for congestion avoidance scenario: a) variations of RTT and RTT_d and b) variations of $e = RTT - RTT_d$.

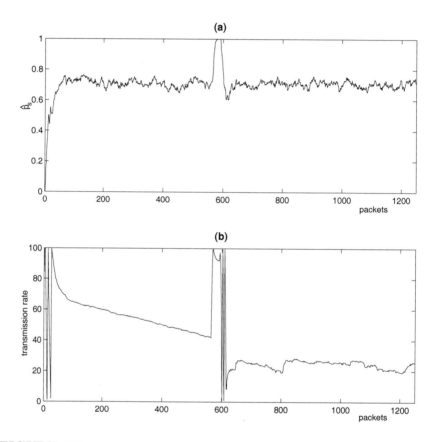

FIGURE 8.7
Picture of the first channel first source for congestion avoidance scenario: a)
variations of $\hat{p}_s$ and b) transmission rate u.

9

NNRC: *Throughput and Fairness Guarantees*

CONTENTS

9.1   Overview

This chapter focuses on the design of the throughput control module of the *NNRC* framework. Initially the necessity of its presence is analyzed and its relationship to fairness in computer networks is clarified. Robustness against modeling imperfections, exogenous disturbances (UDP traffic) and delays (propagation and buffering) are proven. The developed number of channels is guaranteed to be bounded by specific preassigned limits, while modifications are provided to achieve their reduction whenever congestion is detected. Finally, the theoretically developed throughput unit is tested via illustrative simulations studies performed on a small scale network, highlighting various performance measures (i.e., buffer level, source allocation) and verifying theoretical results.

9.2   Necessity for Throughput Control

Let us consider the dummy network topology pictured in Figure 9.1, which consists of a single 10 Mbps bottleneck with 30 ms propagation delay shared by two neural network rate control (*NNRC*) sources. Each source-destination

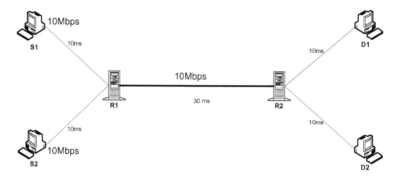

FIGURE 9.1

The dummy network topology.

pair is connected to the bottleneck via a 10 Mbps, 10 ms link and has two way propagation delay equal to 100 ms. All packets are of the same fixed size of 1000 bits and each source (Si, $i = 1, 2$) has to transmit an amount of packets to the corresponding destination (Di, $i = 1, 2$). The number of communication channels per source is assumed constant and equal to 100.

Under these circumstances, each source sends 100,000 bits every round trip time, achieving throughput equal to 1 Mbps. As a consequence, only 2 Mbps out of 10 Mbps are utilized. Such a 20% utilization eventually results in network starvation.

To maximize throughput, all network sources must raise the allocated number of channels from 100 to 500. Clearly, computing this number requires knowledge of the available bandwidth of the bottleneck link and the source round trip time. Unfortunately, no such information is available at the source a priori. Furthermore, owing to the dynamic nature of computer networks, neither may be considered constant throughout transmission. Actually, phenomena like link failures, routing changes, bursts of competing traffic, mobility of wireless links etc., may cause available bandwidth, as well as source RTT, to vary significantly. The aforementioned observations render any thought of pre-computation totally non-applicable. On the other hand, incorporating a worst-case treatment (i.e., allocating to each source the maximum number of channels it supports) is also intractable, as it severely raises implementation complexity.

The above-mentioned arguments clearly establish the necessity of developing algorithms capable of adaptively regulating the number of channels in each source according to local network information.

Additionally, an acceptable congestion control mechanism should possess high levels of TCP friendliness. The latter means that no source should consume an increased amount of network resources when compared to TCP flows. As TCP is the dominant transport protocol in the Internet, if a new protocol acquires an unfair capacity, it will most likely cause network congestion collapse. Within *NNRC*, throughput control is responsible for achieving fair allocation of network resources among competing sources, as well as for adaptively regulating the number of channels in each source.

9.3 Problem Definition

As already mentioned, each source prior to transmission divides the total number of packets into groups. Let us assume that a source s has already transmitted k groups and is about to transmit the k+1 group. Furthermore, recall that with n_s we denote the number of channels associated with source s and with $\overline{RTT}_s$ the mean round trip time achieved at the k-th group. Moreover, it is assumed that $\hat{p}_s$ remains practically constant throughout group transmission.

As we have pointed out in Chapter 8, *NNRC* aims at regulating the per packet round trip time around a piecewise constant RTT_d profile. In achieving such a task, RTT_s becomes practically equal to $\overline{RTT}_s$. Hence, with no loss of generality, the following expression relating source transmission rate (x_s) and number of channels, holds:

$$x_s = \frac{n_s}{\overline{RTT}_s}. \tag{9.1}$$

In each source a utility function $U(x_s)$ of the transmission rate x_s is assigned, satisfying:

- $U(n_s/\overline{RTT}_s)$ is a strictly concave, nondecreasing function of n_s.

- The second derivative (U'') of U with respect to n_s is bounded.

To continue, let us define the objective function $F_o()$ as:

$$F_o(x_s) = U(x_s) - x_s \hat{p}_s. \tag{9.2}$$

Substituting (9.1) into (9.2) and after taking the partial derivative of $F_o()$ with respect to n_s we obtain:

$$\frac{\partial F_o}{\partial n_s} = U'\left(\frac{n_s}{\overline{RTT}_s}\right)\frac{1}{\overline{RTT}_s} - \frac{\hat{p}_s}{\overline{RTT}_s}. \tag{9.3}$$

Define $e_c = U'\left(\frac{n_s}{\overline{RTT}_s}\right)\frac{1}{\overline{RTT}_s} - \frac{\hat{p}_s}{\overline{RTT}_s}$. We will show that if n_s is updated to guarantee $e_c = 0$, then equal sharing of resources among competing sources can be achieved at equilibrium (i.e., $(\hat{p}_s, e_c) = (\hat{p}_s^*, 0)$). Stated otherwise, Jain's fairness index [146] becomes equal to $J = 1$. Notice that the aforementioned procedure also leads to the maximization of (9.2).

Remark 9.1 *In [194], [221] the maximization of the objective function (9.2),* *for a given path congestion measure $\hat{p}_s$, was required to determine source send-ing rate and consequently the equilibrium and the dynamic behavior of the overall system. In our case, the concurrent operation of the* NNRC *algorithm validates (9.1), thus leading through the maximization of (9.2), to a fair op-eration defined via the Jain index.*

Continuing, let us assume for the sake of clarity that we have two com-peting sources sharing the same bottlenecks. It is commonly considered that at equilibrium both will experience practically identical path congestion mea-sures (i.e., $\hat{p}_{s1}^* = \hat{p}_{s2}^*$). Hence, achieving $e_{c1} = e_{c2} = 0$ results in

$$\frac{U'(n_{s1}/\overline{RTT}_{s1})}{U'(n_{s2}/\overline{RTT}_{s2})} = \frac{\hat{p}_{s1}^*}{\hat{p}_{s2}^*} = 1. \tag{9.4}$$

Exploiting the properties of the utility function, (9.4) leads to $\frac{n_{s1}}{RTT_{s1}} = \frac{n_{s2}}{RTT_{s2}}$ and consequently to equal transmission rates at equilibrium, (i.e., $J = 1$).

Summarizing, the problem we try to solve is formulated as follows.

Problem 9.1 *Derive a fully distributed number of channels selection algo-rithm of the form:*

$$\dot{n}_s = H_s(n_s, \hat{p}_s, e_c) \tag{9.5}$$

that operates during group transmission, which when connected to the NNRC *scheme summarized in Chapter 8 guarantees:*

a) *equal sharing of system resources among competing sources at equilibrium $(\hat{p}_s, e_c) = (\hat{p}_s^*, 0)$ (resource allocation property) through maximizing (9.2);*

b) *small enough queues at equilibrium to prevent large queueing delays and congestion collapse, though not zero to avoid link starvation, (network uti-lization property).*

9.4   Throughput Control Design

The purpose of this section, is to design the function H_s in (9.5), to solve the first part of Problem 9.1. The second part, (i.e., the network utilization property) is the main responsibility of the rate controller and thus it will not be further analyzed, as it was treated in Chapter 8. According to the previous section, we have to achieve $e_c = 0$. In this spirit, we differentiate e_c with respect to time to obtain:

$$\dot{e}_c = U''\left(\frac{n_s}{RTT_s}\right)\dot{n}_s - \frac{\dot{\hat{p}}_s}{RTT_s}. \tag{9.6}$$

Notice that in each group transmission, $\overline{RTT}_s$ is kept constant and equal to the value predicted at the beginning of the specific group. Hence, (9.6) is valid.

Let us now define:

$$\dot{n}_s = ae_c n_s + w_c S_c(n_s, \hat{p}_s, e_c, w_c) \tag{9.7}$$

where $a > 0$ is a design constant, $w_c \in \Re$ is a time varying parameter and $S_c(n_s, \hat{p}_s, e_c) \in \Re$ is a nonlinear and bounded scalar function. Guidelines for the construction of $S_c(n_s, \hat{p}_s, e_c)$ shall be given in the subsections that follow.

Consider the Lyapunov function candidate:

$$V = \frac{1}{2}e_c^2 + \frac{1}{2}w_c^2. \tag{9.8}$$

Differentiating (9.8) with respect to time and after substituting (9.7) we obtain

$$
\begin{aligned}
\dot{V} &= ae_c^2 U''(\frac{n_s}{\overline{RTT}_s})n_s - e_c\frac{\dot{\hat{p}}_s}{\overline{RTT}_s} + w_c\dot{w}_c \\
&\quad + e_c U''(\frac{n_s}{\overline{RTT}_s})w_c S_c(n_s, \hat{p}_s, e_c, w_c).
\end{aligned}
\tag{9.9}
$$

Notice that $\dot{\hat{p}}_s$ in (9.9) is provided by (7.2) and thus is considered known. If we choose

$$
\begin{aligned}
\dot{w}_c &= -e_c U''(\frac{n_s}{\overline{RTT}_s})S_c(n_s, \hat{p}_s, e_c, w_c) \\
&\quad + \frac{1}{w_c}e_c\frac{\dot{\hat{p}}_s}{\overline{RTT}_s}
\end{aligned}
\tag{9.10}
$$

$\dot{V}$ becomes:

$$\dot{V} = ae_c^2 U''\left(\frac{n_s}{\overline{RTT}_s}\right)n_s. \tag{9.11}$$

Since $U\left(\frac{n_s}{\overline{RTT}_s}\right)$ is strictly concave, we have $U''\left(\frac{n_s}{\overline{RTT}_s}\right) < 0$ and thus $\dot{V} \leq 0$.

Remark 9.2 *The update law (9.10) is valid as long as $w_c \neq 0$, $\forall t \geq 0$. Such a result may be easily accomplished by employing the well-known projection modification algorithm [130] on (9.10), which moreover does not harm the stability properties established in its absence.*

Following standard arguments from adaptive control literature [137], [103], [247] it can be proved that $\lim_{t \to \infty} e_c(t) = 0$ (provided $n_s \in L_\infty$), while all other signals in the closed loop remain bounded. Thus we state the theorem:

Theorem 9.1 *Consider the error system (9.6). The number of channels selection algorithm (9.7), (9.10) equipped with a projection modification on (9.10) to obtain $w_c \neq 0$, $\forall t \geq 0$ guarantees:*

- $e_c(t), w_c(t) \in L_\infty$

- $\lim_{t \to \infty} e_c(t) = 0$

provided n_s is bounded with $n_s \neq 0$ and $\hat{p}_s < 1$.

The requirement $\hat{p}_s < 1$ is logical, since any resource allocation property is meaningful only at equilibrium, which can only be achieved far from $\hat{p}_s = 1$.

9.4.1 Guaranteeing Specific Bounds on the Number of Channels

Theorem 9.1 is valid provided n_s is bounded away from zero. However, the presented analysis does not provide such boundedness guarantees. In the present subsection, we define the $S_c(n_s, \hat{p}_s, e_c, w_c)$ term in (9.7) accordingly, to establish that the number of channels n_s will vary within certain pre-assigned limits, always under the $\hat{p}_s < 1$ assumption. In other terms, we shall prove that $0 < n_s^{min} \leq n_s \leq n_s^{max}$ with n_s^{min}, n_s^{max} are some known positive integers.

Let us define

$$S_c(n_s, \hat{p}_s, e_c, w_c) = \beta_{c0} S_{c0}(\hat{p}_s, n_s) + \beta_{c1} + \beta_{c2} \tag{9.12}$$

with

$$\beta_{c0} = \begin{cases} 0, & \text{if } (n_s = n_s^{min} \text{ and } M < 0) \\ & \text{or } (n_s = n_s^{max} \text{ and } M > 0) \\ & \text{or } (\hat{p}_s = 1) \\ \\ 1, & \text{otherwise} \end{cases} \tag{9.13}$$

$$\beta_{c1} = \begin{cases} -\frac{ae_c n_s}{w_c}, & \text{if } ((n_s = n_s^{min} \text{ and } M < 0) \\ & \text{and } \hat{p}_s \neq 1) \text{ or } ((n_s = n_s^{max} \\ & \text{and } M > 0) \text{ and } \hat{p}_s \neq 1) \\ \\ 0, & \text{otherwise} \end{cases} \tag{9.14}$$

$$\beta_{c2} = \begin{cases} \frac{-ae_c n_s}{w_c}, & \text{if } (n_s = n_s^{min} \text{and } \hat{p}_s = 1) \\ \\ \frac{-ae_c n_s - \sigma}{w_c}, & \text{if } (n_s > n_s^{min} \text{and } \hat{p}_s = 1) \\ \\ 0, & \text{otherwise} \end{cases} \tag{9.15}$$

with $M = ae_c n_s + w_c$ and $\sigma > 0$ is a design constant.

Assuming that $\hat{p}_s < 1$ and that we start with an initial condition $n_s(0) \in [n_s^{min}, n_s^{max}]$, we distinguish two cases.

Case 1 ($n_s = n_s^{min}$): The number of channels has reached its lower limit. Employing (9.12) - (9.15) into (9.7), it is straightforward to conclude that whenever $n_s = n_s^{min}$ and there is a tendency to move downwards, (i.e., $M < 0$), then $\dot{n}_s = 0$.

Case 2 ($n_s = n_s^{max}$): The number of channels has reached its upper limit. Again using (9.12) - (9.15) we obtain from (9.7) that whenever $n_s = n_s^{max}$ and there is a tendency to move upwards, (i.e., $M > 0$), then $\dot{n}_s = 0$.

Hence, in both cases the constraint $n_s^{min} \le n_s \le n_s^{max}$ is regularly satisfied.

9.4.2 Reducing Channels in Congestion

The analysis presented thus far is valid under the assumption that the source does not detect congestion in its path (i.e., $\hat{p}_s < 1$). Whenever such an assumption is violated, the only acceptable policy would be to reduce the number of channels until $\hat{p}_s < 1$. The purpose of this subsection is to guarantee that $\dot{n}_s \le 0$ whenever $\hat{p}_s = 1$, with the equality ($\dot{n}_s = 0$) to hold only when the number of channels has reached its lower limit ($n_s = n_s^{min}$). To continue, we distinguish the following cases:

Case 1 ($\hat{p}_s = 1$ and $n_s^{min} < n_s \le n_s^{max}$): Employing (9.12) - (9.15) into (9.7) we finally obtain,

$$\dot{n}_s = -\sigma < 0. \tag{9.16}$$

Hence, we conclude that $\dot{n}_s < 0$ whenever $\hat{p}_s = 1$ and $n_s^{min} < n_s < n_s^{max}$.

Case 2 ($\hat{p}_s = 1$ and $n_s = n_s^{min}$): Substituting (9.12) with $b_{c0}(n_s, \hat{p}_s, e_c, w_c) = 0$, $b_{c1}(n_s, \hat{p}_s, e_c, w_c) = 0$ and $b_{c2}(n_s, \hat{p}_s, e_c, w_c) = -\frac{ae_c n_s}{w_c}$ into (9.7) we again obtain $\dot{n}_s = 0$.

Remark 9.3 *Notice that (9.7) outputs a real number in $[n_s^{min}, n_s^{max}]$. However, by definition n_s is a positive integer. To guarantee the aforementioned property, n_s is rounded off to the smallest integer when $e_c < 0$ and to the biggest integer when $e_c > 0$. To explain such a selection, let us denote by d_n the positive (if $e_c > 0$) or negative (if $e_c < 0$) quantity we need to add to (9.7) to achieve the round off. Apparently, $\dot{V}$ is augmented with the extra term $e_c U'' \left(\frac{n_s}{RTT_s} \right) d_n$ which in any case is negative. Hence, the stability properties proven thus far are not harmed owing to the proposed round off procedure.*

The results of the throughput controller design are summarized in Table 9.1.

TABLE 9.1
Throughput controller

Actual System: $F_o(x_s) = U(x_s) - x_s \hat{p}_s$

Error System: $\dot{e}_c = U''\left(\frac{n_s}{RTT_s}\right)\dot{n}_s - \frac{\dot{p}_s}{RTT_s}$

Control Law: $\dot{n}_s = ae_c n_s + w_c S_c(n_s, \hat{p}_s, e_c, w_c)$

Update Law: $\dot{w}_c = -e_c U''\left(\frac{n_s}{RTT_s}\right)S_c(n_s, \hat{p}_s, e_c, w_c) + \frac{1}{w_c}e_c\frac{\dot{p}_s}{RTT_s}$

$S_c(n_s, \hat{p}_s, e_c, w_c) = \beta_{c0}S_{c0}(\hat{p}_s, n_s) + \beta_{c1} + \beta_{c2}$

$\beta_{c0} = \begin{cases} 0, & \text{if } (\hat{p}_s = 1) \text{ or } (n_s = n_s^{min} \text{ and } M < 0) \\ & \text{or } (n_s = n_s^{max} \text{ and } M > 0) \\ 1, & \text{otherwise} \end{cases}$

$\beta_{c1} = \begin{cases} -\frac{ae_c n_s}{w_c}, & \text{if } ((n_s = n_s^{min} \text{ and } M < 0) \\ & \text{and } \hat{p}_s \neq 1) \text{ or } ((n_s = n_s^{max} \\ & \text{and } M > 0) \text{ and } \hat{p}_s \neq 1) \\ 0, & \text{otherwise} \end{cases}$

$\beta_{c2} = \begin{cases} \frac{-ae_c n_s}{w_c}, & \text{if } (n_s = n_s^{min} \text{ and } \hat{p}_s = 1) \\ \frac{-ae_c n_s - \sigma}{w_c}, & \text{if } (n_s > n_s^{min} \text{ and } \hat{p}_s = 1) \\ 0, & \text{otherwise} \end{cases}$

$M = ae_c n_s + w_c, \ \sigma > 0$

Properties: $e_c(t), w_c(t) \in L_\infty, \ \lim_{t\to\infty} e_c(t) = 0, \ 0 < n_s^{min} \leq n_s \leq n_s^{max},$
$\dot{n}_s \leq 0$ whenever $\hat{p}_s = 1$ and $n_s \in (n_s^{min}, \ n_s^{max}),$
$\dot{n}_s = 0$ whenever $\hat{p}_s = 1$ and $n_s = n_s^{min}$

Conditions: $w_c \in \Re, \ S_c(n_s, \hat{p}_s, e_c) \in \Re$

Requirements: $U(n_s/\overline{RTT}_s)$ is a strictly concave, nondecreasing function
and (U'') is bounded.

9.5   Illustrative Example

In this section we evaluate the performance of the proposed throughput control unit by presenting simulations performed in the same simple network used in Chapter 8, altering slightly its parameters as shown in Figure 9.2 to enhance the presentation of results. Each *NNRC* source (Si, $i = 1, ..., 5$) has to transmit 50,000 packets to the corresponding destination (Di, $i = 1, ..., 5$). Additionally, each connection has a queue with maximum capacity (B_{max}) 300 packets.

To illustrate the performance of the proposed algorithm, three simulation scenarios have been conducted. In the first, *NNRC* operation is tested under

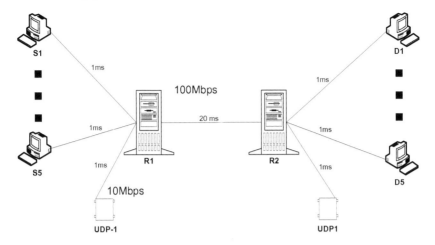

FIGURE 9.2
The single-node network topology.

normal traffic conditions. In the second, bursty traffic is considered and the congestion avoidance mechanism of channel reduction is illustrated. Finally, in the third, throughput improvement is demonstrated when significant packet delay variations owing to non-congestion issues (i.e, re-routing) are witnessed.

9.5.1 Implementation Details

Implementation details regarding both the feasible desired round trip time estimator and the rate control units can be found in the corresponding section of Chapter 8. In this subsection, attention is paid only to the design of the throughput controller. The regressor vector of the unit was chosen as

$$S_{c0}(\hat{p}_s, n_s) = \frac{2}{1 + e^{-(\frac{n_s}{n_s^{max}} - 0.2)}} \left[(\frac{-3}{1 + e^{-30(\hat{p}_s - 0.2)}} + 3) \right.$$
$$\left. (\frac{-2}{1 + e^{-20(\hat{p}_s - 0.55)}} + 1) \quad (\frac{-5}{1 + e^{-20(\hat{p}_s - 0.8)}}) \right]^T.$$

The values of parameters a, σ that appear in (9.7) and (9.15) were $a = 0.95$, $\sigma = 5$.

In (9.7) $n_s(0) = 200$, while the initial weight values were selected as:

$$w_c(0) = [5 \quad 5 \quad 5].$$

Clearly from (9.7), the parameter a regulates the speed of convergence of n_s and through (9.6) it affects the settling time (t_s) of Jain's index (J). The

Jain index is calculated via

$$J = \frac{(\sum_s x_s)^2}{\sum_s \nu x_s^2}$$

for a total number of sources equal to $\nu = 5$ and with x_s as it results from (9.1). The settling time t_s is defined as the time required for Jain Index (J) to enter the 2% zone of $J = 1$ (i.e., $J \geq 0.98, \forall t \geq t_s > 0$).

On the other hand, σ influences the performance of *NNRC* framework only at congestion (i.e., whenever $\hat{p}_s = 1$), as it is related to the amount of channel reduction. Large values of σ will lead to significant changes in the source rate, thus introducing relatively large transients and undesirable oscillations. Smaller values of σ avoid this problem. However, they make the congestion phenomenon more persistent. The aforementioned remarks can be easily verified via simulations.

The utility function was selected as

$$U(\frac{n_s}{RTT_s}) = 100 \ln(\frac{n_s}{RTT_s}).$$

Actually, any utility function $U(\cdot)$ satisfying the properties stated in this chapter can be used. It can be demonstrated through simulations that the parameter that multiplies $ln(\cdot)$ affects the buffer level at equilibrium, as well as the settling time (t_s) of the Jain's index.

9.5.2   Normal Scenario

Within the normal scenario, all sources are operated via *NNRC* and succeed transmitting their packets to the corresponding destinations. The sending times achieved are presented in Table 9.2.

TABLE 9.2
Achieved sending times in the normal scenario

S_1	S_2	S_3	S_4	S_5
30.6917	30.6894	30.6918	30.6919	30.6913

Queue length fluctuations in $R1$–$R2$ connection with respect to time are presented in the Figure 9.3(b). Throughout simulation, utilization of $R1$–$R2$ connection remains, disregarding the transient period, around 100%, as it is depicted in Figure 9.3(a). Fairness is illustrated in Figure 9.3(c), where the Jain index is plotted. After a transient period the Jain index evolves close to unity, which results in almost equal source transmission rates.

We selected the first source ($S1$) to demonstrate the estimate of its path congestion level, the error e_c and the variation of the number of channels. The results appear in Figures 9.4(a),(b) and (c) respectively. Apparently after a

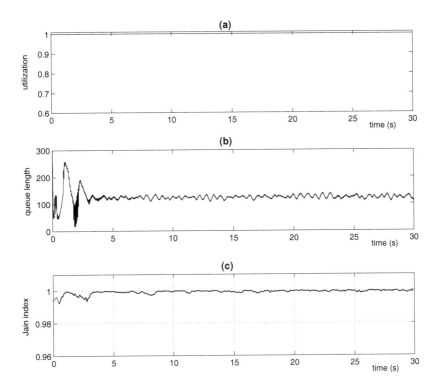

FIGURE 9.3
Performance of *NNRC* framework in the case of a normal scenario: a) utilization of R1–R2 connection, b) queue length of R1–R2 connection, and c) Jain index.

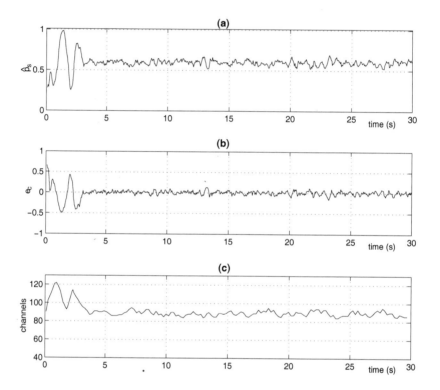

FIGURE 9.4
Performance of source $S1$ in the case of normal scenario: a) variation of $\hat{p}_s$, b) variation of e_c, and c) variation of number of channel (n_s).

period of 3 seconds the source reaches an equilibrium point. The aforementioned results are indicative and can be seen to hold for all other network sources.

9.5.3 Congestion Avoidance Scenario

In this scenario the network is forced to congestion via the abrupt increase of UDP packets. Specifically, the CBR source increases its transmission rate from 10Mbps to 50Mbps. The resulted bursty phenomenon started at t=15s and its duration was 2s.

The sending times achieved by the sources are presented in Table 9.3.

TABLE 9.3
Achieved sending times in congestion avoidance scenario

S_1	S_2	S_3	S_4	S_5
32.2703	32.2691	32.2691	32.2698	32.2704

Critical information regarding the $R1$–$R2$ connection including utilization, queue length variation and fairness expressed via the Jain index are pictured in Figure 9.5

The bursty phenomenon drives the network to congestion followed by a short starvation period. This behavior is clearly observed by studying Figure 9.5(b). The resulted utilization presents a 20% reduction while fairness is preserved (Jain index remains approximately constant around unity) throughout simulation.

To demonstrate the estimate of path congestion level $(\hat{p}_s)$, the control error $e_c = RTT - RTT_d$, as well as the number of channels variation, we plotted the corresponding measurements obtained from source $S1$ (see Figure 9.6). After termination of the bursty phenomenon, all aforementioned measures practically re-establish their level, a fact critical to regain performance.

Concluding *NNRC* presented fast adaptation to sudden network fluctuations, with high levels of fairness among competing sources.

9.5.4 Throughput Improvement

We simulated a scenario where packet delay changes owing to non-congestion issues (i.e., a re-routing scenario), modeled as an increase from 20 ms to 40 ms of the propagation delay of the $R1$–$R2$ connection, started at t=20s.

Performance measures (in the presence of throughput control) related to $R1$–$R2$ connection and specifically utilization, queue length and fairness are plotted in Figure 9.7, while in Figure 9.8 a demonstration of the estimated path congestion level $(\hat{p}_s)$ as well as of the variation of the number of channels allocated are provided through measurements on source $S1$.

Besides the number of channels allocated, which after re-routing attain a

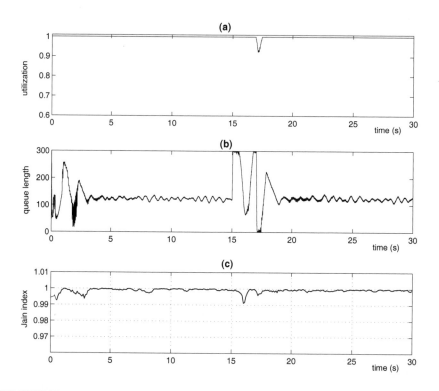

FIGURE 9.5

The performance of *NNRC* framework in the case of congestion avoidance scenario: a) utilization of *R1–R2* connection, b) queue length of *R1–R2* connection, and c) Jain index.

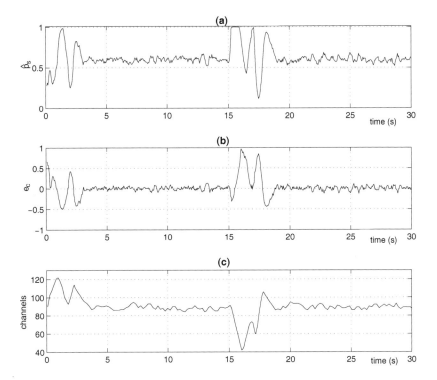

FIGURE 9.6
The performance of source $S1$ in the case of congestion avoidance scenario: a) variation of $\hat{p}_s$, b) variation of e_c, and c) variation of number of channel (n_s).

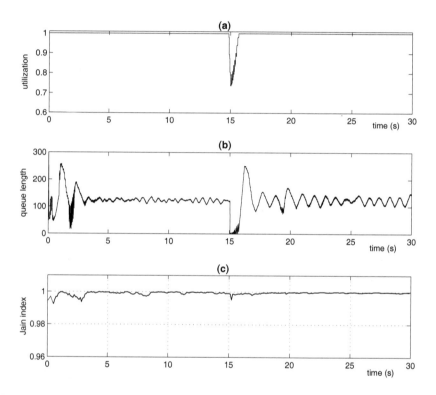

FIGURE 9.7
Performance of *NNRC* framework with throughput control unit (variable number of channels) in case of re-routing scenario: a) utilization of $R1$–$R2$ connection, b) queue length of $R1$–$R2$ connection, and c) Jain index.

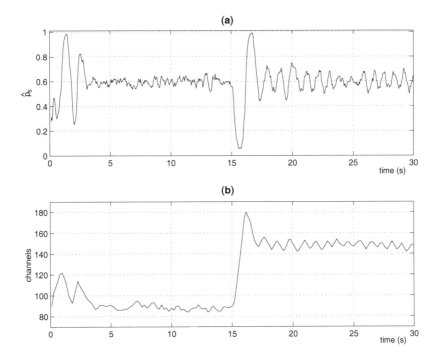

FIGURE 9.8
Performance of *NNRC* framework with throughput control unit (variable number of channels) in case of re-routing scenario: a) variation of $\hat{p}_s$, b) variation of number of channel (n_s).

higher level than before, all other quantities re-established their previously obtained values. The increase in the number of channels is expected, as more channels are required to efficiently recover the increase in propagation delay.

The simulation was repeated keeping this time the number of channels constant at the value of 90. In other terms, the throughput control unit of *NNRC* was de-activated.

The sending times achieved by all sources are summarized in Table 9.4. Significantly smaller sending times were obtained in the presence of throughput control.

Figure 9.9 clearly demonstrates a major contribution of throughput control in improving utilization and queue length of the bottleneck connection $R1$–$R2$. In its absence, the network is driven to a permanent starvation.

The aforementioned results clearly verify that throughput control achieves throughput adaptation, preserving high levels of fairness.

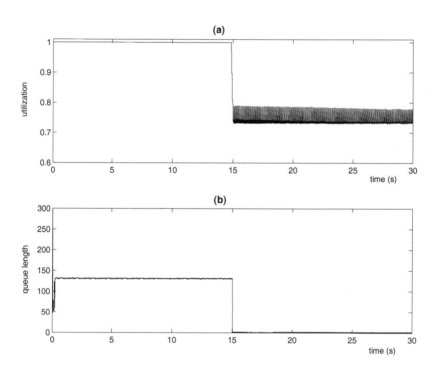

FIGURE 9.9
Performance of *NNRC* framework with constant number of channels ($n_s = 90$) in case of re-routing scenario: a) utilization of $R1$–$R2$ connection, b) queue length variations of $R1$–$R2$ connection. Network starvation is evident.

TABLE 9.4
Achieved sending times in re-routing scenario

	S_1	S_2	S_3	S_4	S_5
with throughput control unit	30.520	30.521	30.520	30.519	30.520
without throughput control unit	36.168	36.168	36.167	36.167	36.167

9.6 Concluding Comments

The chapter considers the design and the rigorous analysis of the throughput control module of the *NNRC* framework. Its necessity and strong relation to fairness is clarified. Robustness against modeling imperfections, exogenous disturbances and delays is also proved. Illustrative simulations studies, performed on a small scale though representative network example, highlight the module's attributes.

Parts of the results presented in this chapter originally appeared in [131]. As in the previous chapter, linear in the weights neural networks has been employed to overcome the vast amount of uncertainties in the system. However, the whole design is not so strongly connected to neural networks. In fact any linear in the parameters function approximation structure (e.g., fuzzy systems, wavelets, polynomials etc.) could be equally used as well.

10

NNRC: *Performance Evaluation*

CONTENTS

10.1 Overview

Our objective within the *NNRC* framework was to develop a source rate control protocol, operated at the source, taking as feedback information regarding network condition and satisfying both a resource allocation property and a network utilization property. These objectives have been outlined in Chapter 9.

In this chapter, we demonstrate, through comparative simulations studies performed on a heterogeneous long-distance high-speed network with multiple bottleneck links, that *NNRC* satisfies the aforementioned properties to a very good extent, outperforming FAST TCP, which is the dominant protocol against all others reported in the relevant literature [293].

To this respect, a series of tests are conducted focusing on a) scalability with respect to variations in network characteristics (i.e., maximum link buffer capacities, propagation delays and link bandwidths); b) dynamic response in the presence of sudden changes of the available bandwidth and propagation delays and c) *NNRC* and FAST TCP interaction (interfairness) when they coexist in a computer network.

In our simulations we mainly consider bulk data transfers (e.g., FTP). However, the performance of the tested protocols is also evaluated in the presence of short web-like flows (e.g., www sources).

10.2   Network Topology

In this section the network topology on which all simulations were realized shall be presented. To increase generalization, efforts have been devoted to utilizing network characteristics leading to representative computer network structures. Specifically, we consider selecting the number of bottleneck connections, the bandwidth and propagation delays, as well as the heterogeneity of sources with respect to RTT.

- **Bottleneck connections:** The number of bottleneck connections in a path is what dominates network operation and not the total number of connections. A bottleneck increases delay and probability of congestion appearance. Technical reports [318] have demonstrated that an internet path is constituted on average by fifteen connections.[1] To introduce the multi-bottleneck characteristic, we consider in all simulations studies the presence of three bottlenecks practically assuming that bottleneck connections constitute 20% of the total path connections.

- **Bandwidth:** The bandwidth of all connections was selected equal to 1 Gbps, a value that is currently the maximum witnessed in computer networks, thus establishing the high-speed character of the topology.

- **Propagation delay:** Considering that a 300 km physical connection corresponds to a propagation delay of 1 ms, to introduce the long-distance property, the propagation delay of each bottleneck connection was selected equal to 50 ms.

- **Heterogeneity:** To achieve the heterogeneity of the sources in the considered network topology we allow the existence of sources with triple RTT.

The heterogeneous multi-bottleneck long-distance high-speed network topology used to conduct our simulations studies is pictured in Figure 10.1. Besides the three 1 Gbps links of 50 ms propagation delay that connect the four routers Ri, $i = 1, ...4$, 21 sources connected to the routers via 1 Gbps independent connections are distinguished. Among them one is a CBR source which persistently submits UDP packets with an average rate of 100 Mbps and 10% variation, thus modeling the network background traffic.

Four sets of source-destination pairs $(Si - Di)$ $i = 1, ...20$ are also distinguished. The first consists of $(Si - Di)$ $i = 1, ..., 5$ with one way propagation delay equal to 152, 156, 161, 166 and 171 ms, respectively. The remainder all have identical one way propagation delay equal to 52 ms. Each source has to transmit 1,000,000 packets to the corresponding destination.

---

[1] In the literature referred to as hops.

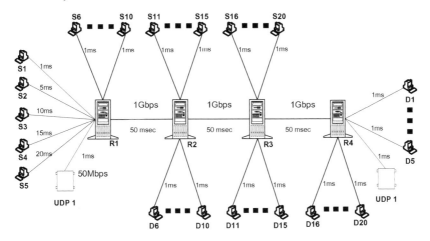

FIGURE 10.1
The heterogeneous, with multiple bottleneck, long-distance high-speed network topology used to conduct the performance analysis.

Moreover all connection buffers have maximum capacity (B_{max}) equal to 3000 packets. All connections mark the entering packets with probability p_m satisfying

$$p_m = \begin{cases} \dfrac{\text{queue length}}{0.7B_{max}} & \text{, if queue length} < 0.7B_{max} \\ \\ 1 & \text{, otherwise} \end{cases} . \qquad (10.1)$$

***NNRC* vs. FAST TCP:** Both *NNRC* and FAST TCP have been implemented and their performance is tested on a number of realistic simulation scenarios, utilizing the network configuration illustrated in Figure 10.1. Specifically, we have investigated: a) their scalability with respect to broad network characteristics variations such as link bandwidth, maximum link capacity, propagation delay; b) their dynamic response against re-routing, bursty traffic, variations in the actual number of users; c) their interaction when they coexist in a computer network.

Throughout the examined simulations studies we assume: (H_1) knowledge of a "good" set of the FAST TCP parameters[2] (α, γ); (H_2) all sources experience empty link buffers at initialization.

To fulfill H_1 some kind of path knowledge is mandatory. Moreover, H_2 is stated to ensure that FAST TCP measures accurately enough the path propagation delay. Clearly, both hypotheses are quite strict and their purpose is to strengthen FAST TCP performance to withstand all comparison tests.

---

[2]For details, the interested reader may consult Chapter 5 where FAST TCP is presented.

At this point we have to underline that no similar assumptions are posed on the *NNRC* side.

10.3   Scalability

It is important for congestion control protocols to be able to maintain their properties as network characteristics change. We thus investigate the scalability of *NNRC* and FAST TCP protocols with respect to variations in maximum link buffer capacity, propagation delays and link bandwidths. When investigating the scalability of the protocols with respect to a particular parameter, we fix the other parameters to the values of the basic setup and we evaluate the performance of the protocol as we change the parameter under investigation.

We conduct our study by considering the multi-bottleneck link network shown in Figure 10.1. In the basic setup, 20 sources share the bottleneck links through access links. The bandwidth of all links in the network is set equal to 1 Gbps, their propagation delay to 50 ms and their maximum link buffer capacity to 3000 packets.

It is considered that the mean rate of the CBR source varies in the range 100 Mbps to 550 Mbps, the propagation delay of $R1$–$R2$ connection in the range 10 ms to 200 ms and the maximum link buffer capacity in the range 200 to 3000 packets. In this study, the following measures have been utilized to quantify protocol performance:

- average queue length at $R1$–$R2$ connection,

- average utilization of $R1$–$R2$ connection,

- settling time (t_s) of Jain index for the first group of sources (Si, $i = 1, ..., 5$).

The average queue length and the average utilization of $R1$–$R2$ connection are calculated over the entire duration of the simulation, thus containing information about the transient behavior of the system as well.

In our simulations we consider persistent FTP sources with packet size equal to 12,000 bits. The simulation time was set equal to 25 seconds, which is sufficient for the network to reach an equilibrium state. Even though it is highly unrealistic, it is assumed that all sources enter the network concurrently, to better study all possible transient effects.

10.3.1   Effect of Maximum Queue Length

We first evaluate the performance of the *NNRC* and FAST TCP protocols as we change the maximum capacity of the connection buffers. We fix the mean

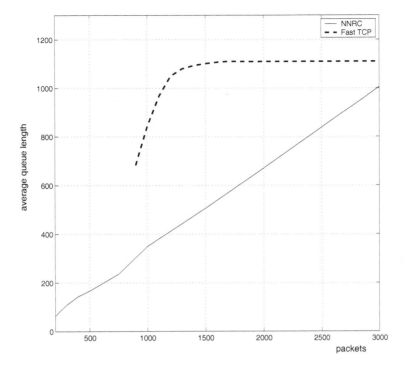

FIGURE 10.2
Average queue length at $R1$–$R2$ connection, with respect to the maximum queue capacity, for *NNRC* and FAST TCP.

rate of CBR source to 100 Mbps, the propagation delays of all links connecting routers Ri, $i = 1, ..., 4$ to 50 ms and their corresponding bandwidths to 1 Gbps and we alter the maximum link buffer capacity (B_{max}) from 200 to 3000 packets.

The average queue length at $R1$–$R2$ connection is presented in Figure 10.2, while Figure 10.3 depicts its average utilization. Finally, the settling time (t_s) of Jain index for the first group of sources $(Si, i = 1, ..., 5)$ is shown in Figure 10.4.

As Figure 10.2 reveals, the efficiency of FAST TCP reduces gradually for maximum link buffer capacity values less than 1500 packets and reaches the point of "no operation" at 900 packets. Evidence of FAST TCP destabilization is the large oscillatory behavior depicted in Figure 10.4 for B_{max} values less than 1100 packets.

On the contrary, *NNRC* is capable of operating efficiently even for $B_{max} = 200$ packets, while the total average queue length curve is kept always below the one achieved by FAST TCP. The linear growth is mostly owing to the fact that B_{max} is directly employed in the derivation of the marking probability

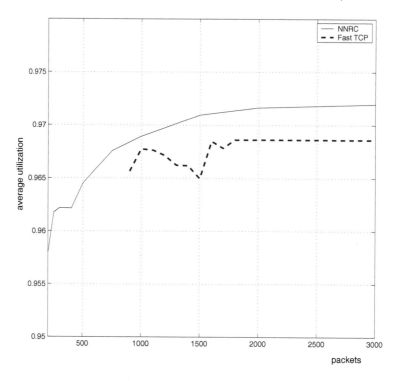

FIGURE 10.3

Average utilization of $R1$–$R2$ connection, with respect to the maximum queue capacity, for *NNRC* and FAST TCP.

(10.1). Qualitatively, an increase in B_{max} results in an increased average queue length equilibrium.

From Figure 10.3 it can be observed that both protocols achieve high utilization (97%) at all B_{max} ranges. Utilization may evolve in the vicinity of 100% if the transient period is excluded. Moreover, as Figure 10.4 clearly illustrates, the duration of transients is small, always below 4 seconds.

10.3.2   Effect of Propagation Delays

We then investigate the performance of *NNRC* and FAST TCP protocols, as we change the propagation delay at $R1$–$R2$ connection, while maintaining all other basic setup characteristics. Any change in the link propagation delay causes a corresponding change in the round trip propagation delay of all sources, which utilize the particular connection.

Average queue length variation at $R1$–$R2$ connection is presented in Figure 10.5, while its average utilization is depicted in Figure 10.6. Finally,

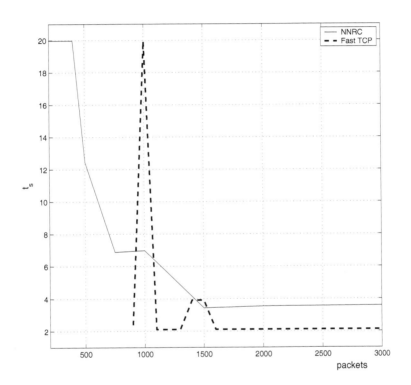

FIGURE 10.4
Average settling time (t_s) of Jain index of the first group of sources $(Si, i = 1, ..., 5)$, with respect to the maximum queue capacity, for *NNRC* and FAST TCP.

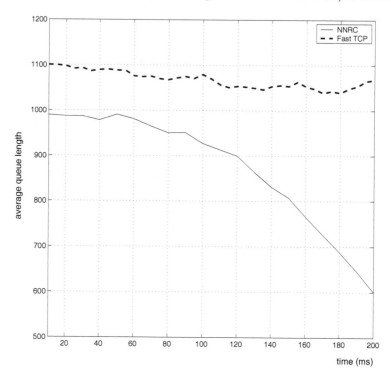

FIGURE 10.5

Average queue length at $R1$–$R2$ connection, with respect to propagation delay of $R1$–$R2$ connection, for $NNRC$ and FAST TCP.

the average settling time (t_s) of Jain index for the first group of sources $(Si,\ i = 1, ..., 5)$ is shown in Figure 10.7.

One of the advantages of $NNRC$ over FAST TCP is its capability to operate with small average queues, which as a result reduces the necessity of utilizing large link buffers. This is evident from studying Figure 10.5.

In addition to low average queues, $NNRC$ achieves high network utilization as Figure 10.6 clearly illustrates. However, this struggle to maintain high utilization levels seems to have a negative impact on average settling time, a phenomenon that appears to be stronger at higher propagation delay values (greater than 150ms).

10.3.3 Effect of Bandwidth

We finally investigate the performance of $NNRC$ and FAST TCP protocols, as we change the link bandwidth. Specifically, we fix the propagation delays of the three links that connect routers $Ri,\ i = 1, ..., 4$ to 50 ms, their bandwidths to 1 Gbps, the maximum link buffers capacity to 3000 packets, and we consider

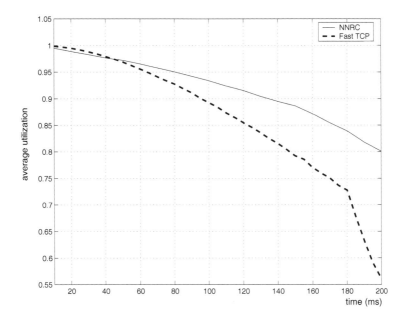

FIGURE 10.6
Average utilization of $R1-R2$ connection, with respect to propagation delay of $R1-R2$ connection, for *NNRC* and FAST TCP.

altering the average rate of the CBR source in the interval [100 Mbps, 550 Mbps]. In that respect, the available link bandwidth is reduced from 900 Mbps to 450 Mbps.

It is further allowed a 10% variation on the average rate of the CBR source, which has a significant impact on performance.

Average queue length variation at $R1-R2$ connection is presented in Figure 10.8, while Figure 10.9 pictures its average utilization. Finally, the average settling time (t_s) of Jain index for the first group of sources $(Si, i = 1, ..., 5)$ is shown in Figure 10.10.

Clearly from Figure 10.9, both *NNRC* and FAST TCP achieve high network utilization $(> 97\%)$ at all bandwidths. However, *NNRC* results in lower average queues, which is especially true as the average CBR source rate attains larger values (see Figure 10.8).

Apparently, FAST TCP outperforms *NNRC* with respect to average settling time, as is clearly illustrated in Figure 10.10. *NNRC* requires more time to achieve fairness, presenting significantly slower convergence to equilibrium as the average CBR source rate exceeds 450 Mbps.

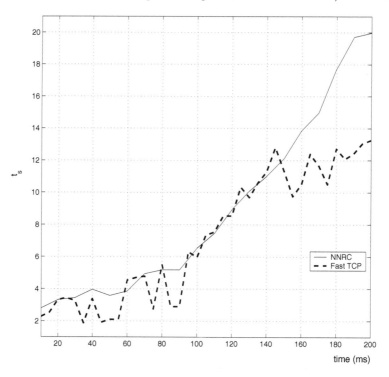

FIGURE 10.7
Average settling time (t_s) of Jain index for the first group of sources $(Si,\ i = 1, ..., 5)$, with respect to propagation delay of $R1$–$R2$ connection, for *NNRC* and FAST TCP.

10.4   Dynamic Response of *NNRC* and FAST TCP

To fully characterize the performance of the protocols, apart from the properties of the system at equilibrium, we need to investigate their transient behavior. The protocols must generate smooth responses which are well damped and converge fast to the desired equilibrium state. To evaluate the transient behavior of *NNRC* and FAST TCP protocols, we consider the multi-bottleneck link network shown in Figure 10.1 and we examine their performance in sudden changes of the available bandwidth and propagation delays. Our principal aim is to examine the responsiveness of the protocols, i.e., their ability to respond quickly without serious oscillations to sudden changes in bandwidth and propagation delays. Such changes can be observed due to a number of reasons:

- link failures,

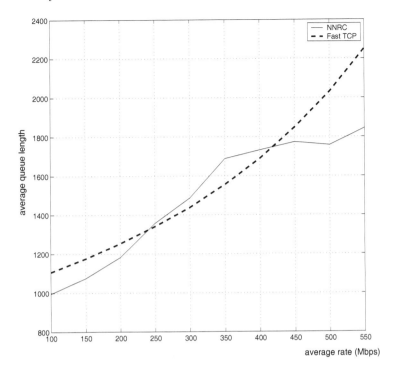

FIGURE 10.8
Average queue length at $R1-R2$ connection, with respect to the average rate
of the CBR source, for *NNRC* and FAST TCP.

- variation of number of sources utilizing the network,

- routing changes,

- bursts of competing traffic,

- mobility of wireless links.

To investigate the dynamic behavior of *NNRC* and FAST TCP protocols
we materialized three scenarios, focusing on the effects caused by bursty traffic, re-routing and non-constant number of sources, respectively. To quantify
performance, the following measures are employed:

- average queue length at $R1-R2$ connection,

- average utilization of $R1-R2$ connection,

- average Jain index for the first group of sources (Si, $i = 1, ..., 5$).

To derive the aforementioned metrics, multiple runs (specifically 10) of
the same scenario have been implemented. The simulations time was kept

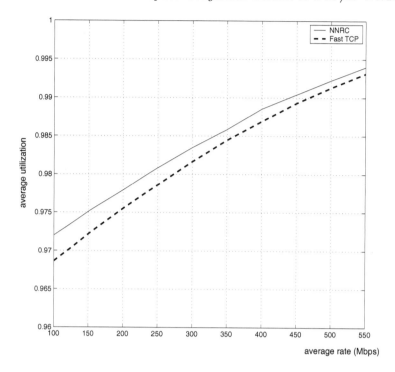

FIGURE 10.9
Average utilization of $R1$–$R2$ connection, with respect to the average rate of
the CBR source, for *NNRC* and FAST TCP.

constant and equal to 50 seconds, which is sufficient for the network to reach
an equilibrium.

10.4.1   Bursty Traffic

The performance of *NNRC* and FAST TCP is studied in the presence of
abrupt variations of the available bandwidth. Such a phenomenon called
bursty traffic may appear in a network owing to the sudden appearance of
UDP sources which are not equipped with congestion avoidance mechanisms.

In the multi-bottleneck network configuration of Figure 10.1, we simulated
a bursty traffic of 10 seconds duration starting at t=20s, modeled as an instantaneous increase of the average rate of the CBR source from 100 Mbps to
300 Mbps.

The average queue length variation at $R1$–$R2$ connection is presented in
Figure 10.11, while the resulted average utilization is pictured in Figure 10.12.
Furthermore, the average Jain index of the first group of sources (Si, $i =$
$1, ..., 5$) is demonstrated in Figure 10.13.

Figure 10.11 clearly reveals that FAST TCP adapts faster with respect

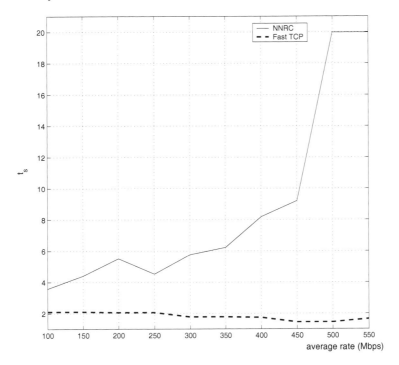

FIGURE 10.10
Average settling time (t_s) of Jain index of the first group of sources $(Si, i = 1, ..., 5)$, with respect to the average rate of the CBR source, for *NNRC* and FAST TCP.

to *NNRC*. Furthermore, both FAST TCP and *NNRC* achieve almost perfect utilization (see Figure 10.12). With respect to fairness, *NNRC* as well as FAST TCP present an almost identically very good behavior, as illustrated in Figure 10.13, with a slight lead of *NNRC* at steady state.

10.4.2 Re-Routing

Owing to link failures, insufficient link available bandwidth etc., a packet may be forced to change its preassigned path to destination. Such a phenomenon, called re-routing, causes abrupt propagation delay variations with effect on performance. To study re-routings, we simulated a sudden decrease of the propagation delay at $R1$–$R2$ connection from 50 ms to 20 ms. We further assume that its duration is 10s, starting at t=20s.

The average queue length variation at $R1$–$R2$ connection is presented in Figure 10.14 while the resulted average utilization is pictured in Figure 10.15. Moreover, the average Jain index of the first group of sources $(Si, i = 1, ..., 5)$ is demonstrated in Figure 10.16.

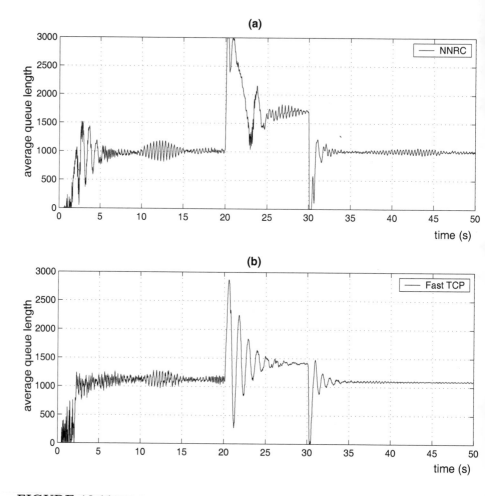

FIGURE 10.11
Average queue length at $R1-R2$ connection, in the presence of bursty traffic, for *NNRC* and FAST TCP. The burst starts at t=20s and its duration is 10s.

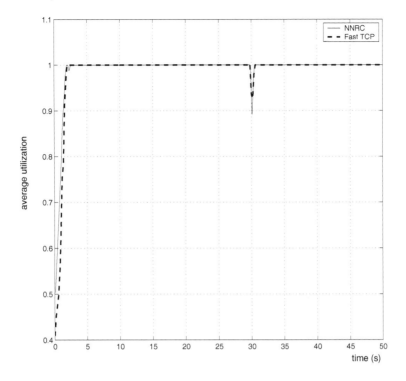

FIGURE 10.12

Average utilization of $R1$–$R2$ connection, in the presence of bursty traffic, for *NNRC* and FAST TCP. The burst starts at t=20s and its duration is 10s.

Apparently, FAST TCP is incapable of handling re-routings as it results in network starvation (see Figure 10.14), with poor utilization as a direct consequence (see Figure 10.15). On the contrary, *NNRC* recovers fast while maintaining high levels of network utilization. With respect to fairness, both FAST TCP and *NNRC* present an almost identically very good behavior as illustrated in Figure 10.16 with a slight lead of *NNRC* at steady state.

The starvation observed in the re-routing scenario brings to the surface a major drawback of FAST TCP connected to the precise propagation delay measurement assumption.

In all FAST TCP simulation studies presented thus far, propagation delay results from the minimum RTT measured from the beginning of packet transmission. Such a measurement is precise indeed, assuming initially that all network link buffers are empty (i.e., zero queueing delays). Clearly, in the presence of a re-routing, leading to propagation delay reduction, the aforementioned assumption is severely violated.

To address this issue, we executed the modified version of FAST TCP proposed in [293] according to which the minimum RTT observed in a time

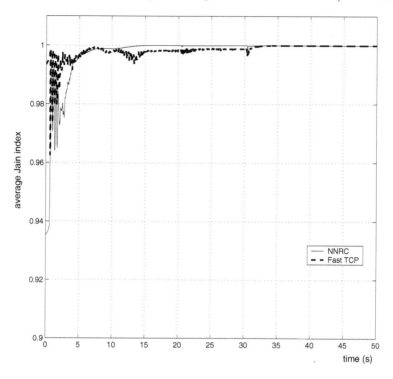

FIGURE 10.13
Average Jain index of the first group of sources $(Si,\ i = 1, ..., 5)$, in the presence of bursty traffic, for *NNRC* and FAST TCP. The burst starts at t=20s and its duration is 10s.

window is used as a candidate propagation delay measurement. We simulated the modified FAST TCP and results concerning the average queue length at $R1$–$R2$ connection, its average utilization and Jain index of the first group of sources $(Si,\ i = 1, ..., 5)$ are demonstrated in Figures 10.17-10.19, respectively. Besides the apparent improvement, *NNRC* still outperforms FAST TCP.

10.4.3 Non-Constant Number of Sources

In realistic network conditions, the number of sources utilizing its links is never constant. Instead, dynamic variations are commonly witnessed. Such a behavior affects the available link bandwidth and consequently the performance of every congestion control protocol. To simulate the phenomenon, we consider the basic network configuration of Figure 10.1. However, now, at t=20s sources $S4$, $S5$, which belong to the first group $(Si,\ i = 1, ..., 5)$, stop transmission, reducing the total number of sources of the first group utilizing the network to 3. At t=30s sources $S4$, $S5$ resume transmission.

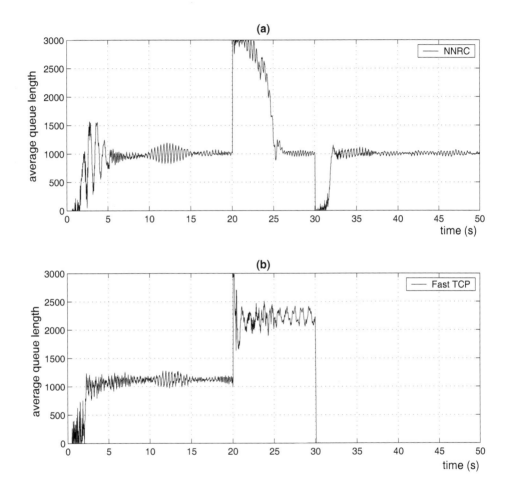

FIGURE 10.14

Average queue length at $R1$–$R2$ connection, in the presence of re-routing, for *NNRC* and FAST TCP. The re-routing appears at t=20s and its duration is 10s.

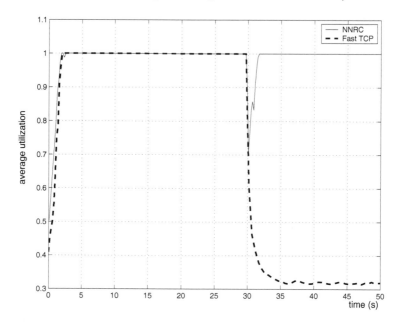

FIGURE 10.15

Average utilization of $R1$–$R2$ connection, in the presence of re-routing, for *NNRC* and FAST TCP. The re-routing starts at t=20s and its duration is 10s.

The average queue length variation at $R1$–$R2$ connection is presented in Figure 10.20, while the resulted average utilization is pictured in Figure 10.21. The average Jain index of the first group of sources (Si, $i = 1, ..., 5$) is demonstrated in Figure 10.22.

Studying Figures 10.20, 10.21 it is concluded that FAST TCP reacts more smoothly to the variation in the number of sources, achieving a slightly improved utilization. Its main drawback though is on fairness. As is evident from Figure 10.22, at t=30s when both $S4$, $S5$ resume transmission, an almost 30% reduction on the average Jain index is observed.

10.5 *NNRC* and FAST TCP Interfairness

In the comparison studies presented thus far, it was assumed that all sources were operated under the same protocol (i.e., either FAST TCP or *NNRC*). However, in real life, the expectation of satisfying such an assumption is low, owing to the large number of sources utilizing the network and the variety of developed congestion control protocols. In that respect, it is mandatory

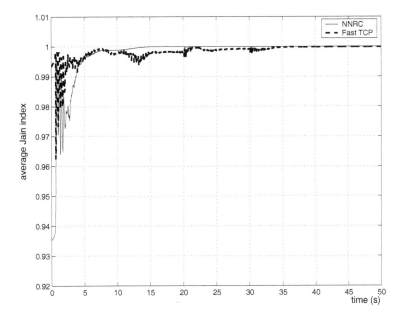

FIGURE 10.16
Average Jain index of the first group of sources $(Si, \ i = 1, ..., 5)$, in the presence of re-routing, for *NNRC* and FAST TCP. The re-routing starts at t=20s and its duration is 10s.

to investigate how FAST TCP and *NNRC* interact when they coexist in a computer network.

For that purpose the basic setup of the multi-bottleneck heterogeneous network configuration presented in Figure 10.1 is employed. However this time, the protocols used in all sources of the four groups formed are not identical.

We vary the number of *NNRC* sources in each group from one to four (and the number of FAST TCP sources is hence varied from four to one).

The average queue length variation at *R1–R2* connection is presented in Figure 10.23, while the resulted average utilization is pictured in Figure 10.24. The average Jain index of the first group of sources $(Si, \ i = 1, ..., 5)$ is demonstrated in Figure 10.25.

Studying the average queue plots, it is evident that improvement in network performance is likely to happen, as *NNRC* dominates. On the contrary, both utilization and fairness standards are kept high in all simulations.

The general conclusion is that FAST TCP and *NNRC* can coexist in a computer network, with the *NNRC* to be less influenced by the presence of FAST TCP.

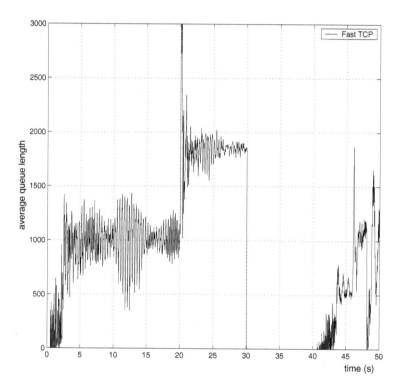

FIGURE 10.17
Average queue length at $R1$–$R2$ connection, in the presence of re-routing, for the modified for FAST TCP protocol.

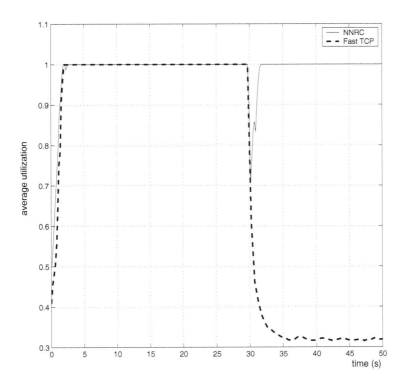

FIGURE 10.18
Average utilization of $R1$–$R2$ connection, in the presence of re-routing, for the modified FAST TCP protocol.

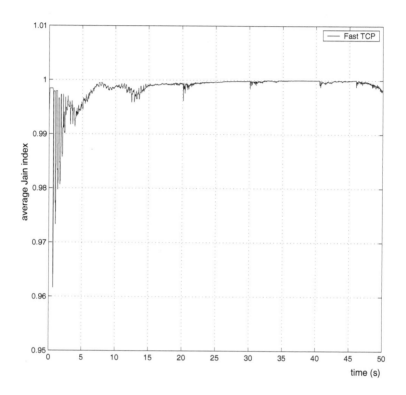

FIGURE 10.19
Average Jain index of the first group of sources $(Si, \; i = 1, ..., 5)$, in the presence of re-routing, for the modified FAST TCP protocol.

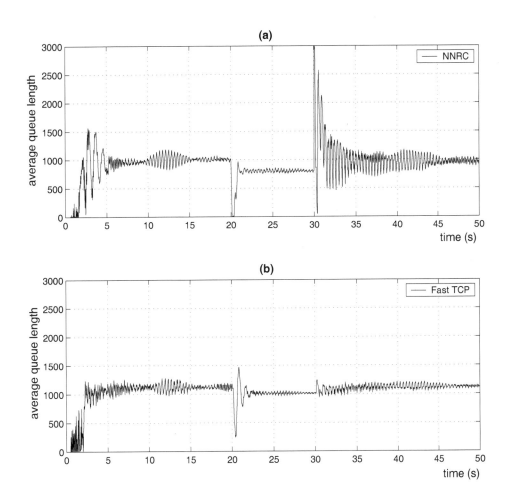

FIGURE 10.20
Average queue length at $R1$–$R2$ connection, in the presence of dynamically varying number of sources.

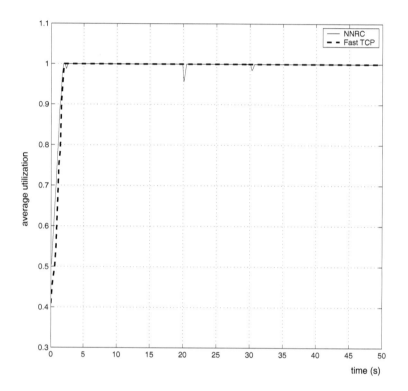

FIGURE 10.21

Average utilization of R1–R2 connection, in the presence of dynamically vary-ing number of sources.

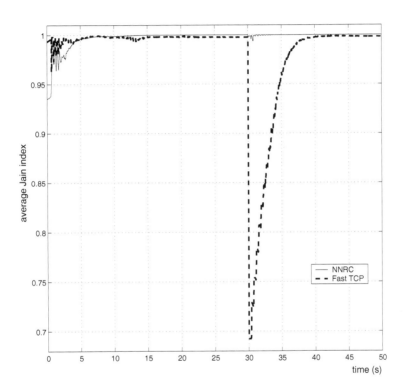

FIGURE 10.22
Average Jain index of the first group of sources $(Si, \; i \; = \; 1, ..., 5)$, in the presence of dynamically varying number of sources.

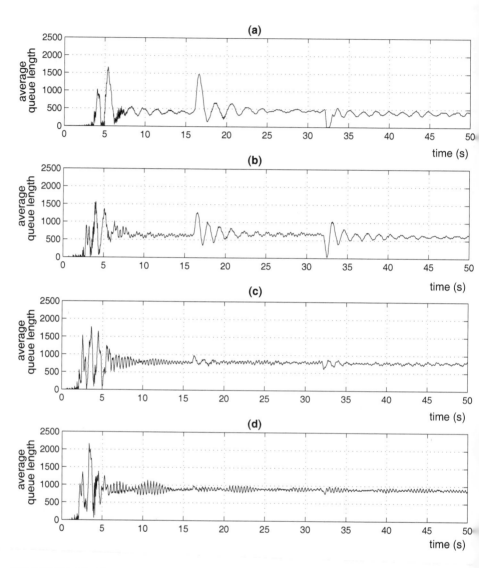

FIGURE 10.23
Average queue length at $R1$–$R2$ connection: a) 1 *NNRC* source and 4 FAST
TCP sources, b) 2 *NNRC* sources and 3 FAST TCP sources, c) 3 *NNRC*
sources and 2 FAST TCP sources, and d) 4 *NNRC* sources and 1 FAST TCP
source.

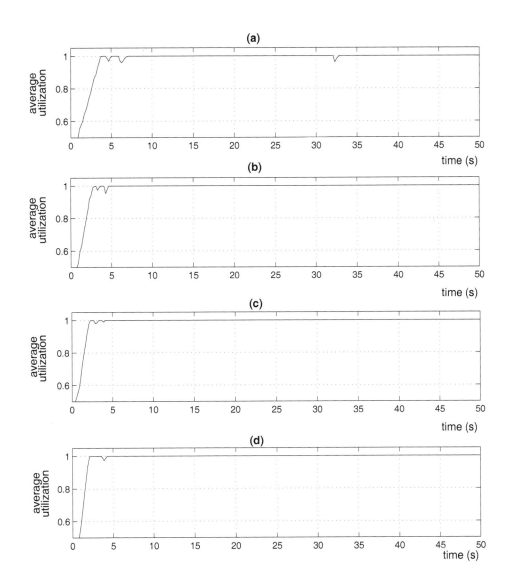

FIGURE 10.24
Average utilization of $R1$–$R2$ connection: a) 1 *NNRC* source and 4 FAST TCP
sources, b) 2 *NNRC* sources and 3 FAST TCP sources, c) 3 *NNRC* sources
and 2 FAST TCP sources, and d) 4 *NNRC* sources and 1 FAST TCP source.

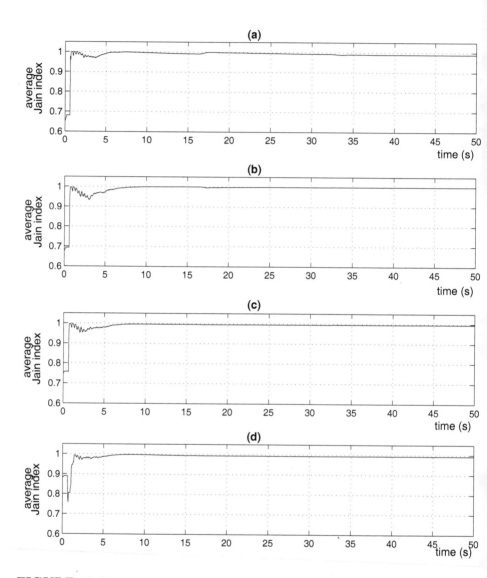

FIGURE 10.25
Average Jain index of the first group of sources $(Si,\ i = 1, ..., 5)$: a) 1 *NNRC* source and 4 FAST TCP sources, b) 2 *NNRC* sources and 3 FAST TCP sources, c) 3 *NNRC* sources and 2 FAST TCP sources, and d) 4 *NNRC* sources and 1 FAST TCP source.

TABLE 10.1
Scalability

	Queue Length Index	Utilization Index	Jain Index
Queue length	*NNRC+*	*NNRC+*	*NNRC+*
Propagation delay	*NNRC*	*NNRC+*	FAST
Bandwidth	$\sim$	*NNRC*	FAST+

10.6 Synopsis of Results

In this section the results obtained thus far are summarized, categorized and presented in a unified manner to aid generalization. Each protocol was tested on its scalability and dynamic response. In case of coexistence, interfairness was also studied.

For the latter, investigations showed that both *NNRC* and FAST TCP can coexist, keeping utilization and fairness at high levels with the *NNRC*, though to be less influenced (in terms of queue length) by the presence of FAST TCP.

Scalability and dynamic response results are concisely summarized in Tables 10.1 and 10.2, respectively.

To indicate superiority of a protocol we put the (+) symbol next to its name. Slight lead of a protocol with respect to the other is indicated by simply stating its name, while the symbol ($\sim$) is used to notify practically identical performance.

The scalability table clearly shows that *NNRC* scales better than FAST TCP with the disadvantage of enlarging the transient period as the available link bandwidth reduces, which consequently leads to reducing fairness (Jain index).

The dynamic response of *NNRC* is clearly superior to FAST TCP mainly owing to the low quality of the latter in the presence of re-routings.

Concluding, *NNRC* appears to operate significantly better than FAST TCP in almost all aspects studied.

To the aforementioned conclusion we underline that FAST TCP was additionally helped by boosting its performance by a) assuring whenever possible the precise propagation delay measuring assumption; b) employing inside information concerning the network structure to optimize the selection of FAST TCP design parameters (α, γ); c) incorporating modifications to smooth the effects of re-routing.

TABLE 10.2
Dynamic response

	Queue Length Index	Utilization Index	Jain Index
Bursty traffic	FAST	FAST	*NNRC*
Re-routing	*NNRC+*	*NNRC+*	*NNRC*
Variable number of sources	FAST	FAST	*NNRC+*

10.7   Concluding Comments

Extensive simulations presented in this chapter clearly show that the proposed adaptive congestion control framework *NNRC* is able to lead the network in stable equilibrium, characterized by fairness, high utilization, small queues, without experiencing significant packet losses. In addition, the scalability of *NNRC* owing to changes in bandwidth, propagation delays and queue capacities was demonstrated. It was further verified that *NNRC* provides interesting transient properties achieving smooth responses without significant oscillations and fast convergence. Besides its practical importance, this chapter also demonstrates the effectiveness of applying control theory and especially adaptive control techniques to provide valid solutions to a very complex system, such as the Internet.

Despite the aforementioned advantages, a number of interesting problems still remain open for future research, the most important of which are summarized below:

- One limitation of the proposed theory seems to be the incorporation of the *ECN bit* to estimate the network congestion status. Towards this direction the prospect of moving towards the inclusion of additional information e.g., packet loss, time behavior of the per packet *RTT*, seems interesting.

- *NNRC* parameters selection was carried out through simulations performed on experimental networks incorporating engineering experience and intuition. Their automatic selection, probably with the help of a specifically designed software tool, will greatly enhance the design process.

- *NNRC* operates towards regulating delay which is certainly an application oriented specification. Incorporating additional quality of application measures (e.g., jitter) will significantly increase both its theoretical and application prospects.

- Implementing real-time applications in modern heterogeneous networks requires the achievement of additional strict quality criteria. Enabling the adjustment of qualitative characteristics of real-time applications, depending on network load, would significantly extend the applicability of the *NNRC* framework.

11

User QoS Adaptive Control

CONTENTS

11.1 Overview

The rapid development of interactive multimedia applications over the Internet (i.e., VoIP, P2PTV, Video on Demand, etc.) imposes increasingly elevated demands both on devices and network infrastructure, in order to provide end-to-end Quality of Service (QoS) even in congested network conditions.

The quality performance of a service or an application is a composition of certain QoS characteristics, relative not only to the network resources but to the specific application as well. For example according to [267] servicing reliably a VoIP traffic requires packet loss ratio no more than 1%, one-way latency not exceeding 150ms and average one-way jitter less than 30ms. The achievement of such time-sensitive constraints makes imperative the deployment of traffic management tools (i.e., congestion controllers) capable of supporting such a task, operating within the dynamically varying, highly complex and heterogeneous Internet environment.

On the other hand, end users perceive QoS in a subjective and relative manner that is not always easily quantifiable and is heavily dependent on the application. For example, typically users expect a voice call on a standard phone to be of higher quality compared to a voice call received from a cell phone, regardless of the fact that both calls might be supported by the same network. In addition, end users perceive QoS according to previous behaviors obtained on similar devices and cost of service. The latter is uncorrelated to the actual network capabilities and supportive infrastructure.

Therefore, to maximize user satisfaction, multimedia services have to be customizable, achieving high adaptivity to the the user's needs and to resource availability.

Many applications may adapt to variations in network operating conditions owing to their inherent adaptation capabilities. For example, the video compression standard MPEG [9], [85], [292] can generate variable rate video streams by modifying the I, B, P frame characteristics.

The underlying methodology that provides increased QoS levels from an end user standpoint is application adaptation and as such it has received much attention from the research community. As its name reveals, application adaptation manages QoS at the application level rather than at the lower network level. It adjusts the application requirements to meet the network induced constraints, without significantly altering (if possible) user satisfaction.

For example, in a teleconference application, if the network throughput is degraded, then the communicated video quality may be degraded as well (i.e., less resolution, lower frame rate, less color depth). However, voice and data transfer, which are key ingredients for a successful teleconference, may maintain their high quality i.e., clear sound, high-speed transferring of documents etc.

In this chapter the time-sensitive congestion control framework *NNRC* that was designed to provide guarantees on latency is combined with an application adaptation scheme that first appeared in [246], [245] and that is based on dynamic neural networks and adaptive control techniques, to propose an integrated solution to the underlying problem.

11.2   Application Adaptation Architecture

Application adaptation can be visualized as the composition of two highly coupled problems. The first, refers to mapping QoS and user satisfaction into application and media related parameters. The second is to construct a control mechanism for altering these parameters to provide the user with the expected QoS, always fulfilling the network and application imposed constraints.

11.2.1   QoS Mapping

Mapping media parameters to application QoS [65], [133], [138], [200], [215], [258] is a highly complicated problem typically related individually to the quality of each media involved. Actually, QoS is the collective effect of service performances which determine in a non-trivial way the degree of satisfaction of a user with a service. Moreover, the QoS a user experiences is directly related to the QoS characteristics of the application delivered. For example, a user may be still satisfied if despite a degradation in a music-video quality he is content with excellent sound.

Mapping between user satisfaction and QoS parameters may be achieved by combining the satisfaction function and the set of constraints. To determine whether a parameter will be considered in the satisfaction function or in the system constraints, it depends upon the user, the application and the network infrastructure.

Generally, for a specific bandwidth availability, variations in the size or the rate of the information lead to changes in the performance metrics. However, if we go below some certain lower bounds of performance characteristics then the corresponding multimedia information becomes unacceptable. These values comprise the lower bounds of the system constraints.

Therefore, application parameters that can be modified directly (i.e., media format characteristics) are incorporated in the user satisfaction function, while network characteristics i.e., delay, packet loss rate, jitter etc. form the network constraints that the application adaptation architecture has to guarantee.

Following the application adaptation formulation [246], the user satisfaction function is considered known in QoS management. It is therefore either given or a priori determined. In the latter alternative, a function approximation methodology using high order neural networks (HONNs) was adopted in [246]. Along these lines, it is assumed that in a preliminary phase the user receives off-line samples of multimedia content obtained with different media configurations and characteristics, responding with conveying its satisfaction level, thus forming a set of input-output measurements. Exploiting the approximation capabilities of HONNs, such a measurement set can be used to efficiently derive an approximation of the user satisfaction function. Detailed description of the aforementioned methodology can be found in [246]. In what follows, the user satisfaction function shall be considered known.

11.2.2 Application QoS Control Design

The purpose of the application QoS controller is to determine the necessary values of the media parameters (characteristics) to achieve the user required satisfaction level, while preserving all network operational constraints i.e., available bandwidth, delay, jitter, cost etc.

Let $F(y)$ be a known, positive, smooth and monotonically increasing function denoting the user satisfaction function, with y being a n-dimensional real vector having as elements y_i, $i = 1, 2, ..., n$ that represent the percentage of the maximum allowable value of the corresponding media characteristic, assigned to the service. Thus $0 \leq y_i \leq 1$, $i = 1, 2, ..., n$. Notice, however, that in practice $y_i \neq 0$, $\forall\ i = 1, 2, ..., n$. Hence, it is reasonable to assume that $\epsilon \leq y_i \leq 1$, $i = 1, 2, .., n$, where ϵ is a small positive constant.

The property of $F(y)$ being monotone increasing agrees with the expected user behavior, that increasing values of media characteristics (i.e., frame size, resolution etc.) increases user satisfaction. On the other hand, there exist media characteristics, such as frame rate, whose increment above a certain threshold degrades user satisfaction. However, such cases can be avoided by allowing the maximum of the corresponding media characteristic to appear where the maximum user satisfaction occurs.

Let us further define the network constraints $N_i(y)$, $i = 1, 2, ..., p$ as positive, smooth and monotone increasing functions of y, satisfying $0 < N_i(y) < N_{imax}(t)$, $i = 1, 2, ..., p$.

Such constraints are the available bandwidth, delay, synchronization, cost and so on. In this work, we assume, with no loss of generality, knowledge of the expression of all $N_i(y)$. Moreover, the allowable constraints N_{imax} are generally time-varying and are considered measurable.

The controller regulates on-line the media characteristics y so the user satisfaction $F(y)$ is always close to a desired user satisfaction level F_{min} without violating the network constraints N_{imax} $(i = 1, 2, ..., p)$.

In that respect the system should adjust the media characteristics to drive user satisfaction error close to zero. The error is defined as

$$e = F(y) - F_{min}. \tag{11.1}$$

In [246], the following application QoS controller is proposed:

$$\dot{y}_i = \begin{cases} f_i(y, N) & , \text{ if } \epsilon \leq y_i \leq 1 \\ 0 & , \text{ otherwise} \end{cases} \tag{11.2}$$

with $i = 1, 2, ..., n$, where

$$f_i(y, N) = -aey_i + W_i^T S_i(y, N) \tag{11.3}$$

and $a > 0$ is a design constant.

Moreover, the term $S_i(y, N)$ in (11.3) is defined as

$$S_i(y, N) = S_i(y)S(Q)b_i(y, e) \tag{11.4}$$

where $N = [N_1 \ N_2...N_p]^T$, and Q is the continuous function

$$Q = min\{S(N_{1max}(t) - N_1(y)), S(N_{2max}(t) \\ -N_2(y)), ..., S(N_{pmax}(t) - N_p(y))\} \tag{11.5}$$

with $S(N_{imax}(t) - N_i(y))$, $i = 1, 2, ..., p$, and $S(Q)$ sigmoid functions of the form

$$S(N_{imax}(t) - N_i(y)) = \frac{2}{1 - e^{N_{imax}(t) - N_i(y)}} - 1 \tag{11.6}$$

$$S(Q) = \frac{2}{1 - e^Q} - 1. \tag{11.7}$$

Moreover,

$$b_i(y, e) = \begin{cases} 1 & , \text{ if } Q \geq 0 \\ \frac{aey_i}{W_i^T S_i(y)S(Q)} + \sigma & , \text{ otherwise} \end{cases} \tag{11.8}$$

for all $i = 1, 2, ..., n$ and σ is a positive constant. Furthermore, the $S(N_{imax}(t) - N_i(y))$ is constructed to follow the sign of its argument:

$$S(N_{imax}(t) - N_i(y)) = \begin{cases} < 0 & , \text{ if } N_{imax}(t) - N_i(y) < 0 \\ = 0 & , \text{ if } N_{imax}(t) - N_i(y) = 0 \\ > 0 & , \text{ if } N_{imax}(t) - N_i(y) > 0 \end{cases} \quad (11.9)$$

The application QoS controller (11.1) - (11.9), together with the weights update law

$$\dot{W}_i = -\gamma W_i - e\frac{\partial F(y)}{\partial y_i} S_i(y, N), \ i = 1, 2, ..., n \quad (11.10)$$

and the user satisfaction function satisfying:

$$\frac{\partial F(y)}{\partial y_i} y_i \geq F_0(y) > 0 \quad (11.11)$$

proves the following theorem:

Theorem 11.1 *[246] The controller (11.1)-(11.10) and satisfaction functions $F(y)$ obeying (11.11) guarantee:*

>   *1. In case N_{imax} ($i = 1, 2, ..., p$) are constant, and F_{min} can be reached under the available N_{imax}, (11.2) outputs the required values of media characteristics y_i ($i = 1, 2, ..., n$) that are necessary to achieve regulation to zero of the user satisfaction error, while all $N_i(y)$ is bounded from above by N_{imax}, $\forall t \geq 0$, provided that we start from inside the set $\mathcal{N} = \{y \in \Re^n : N_i(y) < N_{imax}, \forall i = 1, 2, ..., p\}$.*

>   *2. If N_{imax} is not a constant and/or F_{min} cannot be reached under the available N_{imax}, then y is proved uniformly ultimately bounded with respect to the set $\mathcal{N} = \{y \in \Re^n : N_i(y) < N_{imax}, \forall i = 1, 2, ..., p\}$ while e remains bounded and, moreover, decreases as long as $y \in \mathcal{N}$.*

Certain modifications on (11.10) have been provided in [246] to guarantee that all media characteristics y_i remain within $[\epsilon, 1]$.

11.3 *NNRC* Source Enhanced with Application Adaptation

The layout of a *NNRC* source enhanced with application adaptation is illustrated in Figure 11.1.

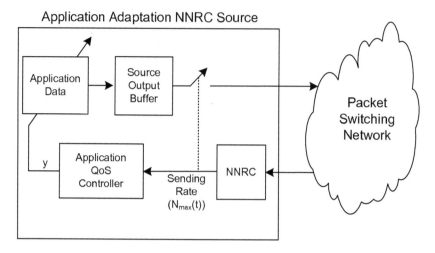

FIGURE 11.1
The layout of a *NNRC* source enhanced with application adaptation.

The *NNRC* congestion controller communicates the calculated sending rate to the application QoS controller, forming the available bandwidth constraint. Consequently, the application QoS controller regulates the media parameters y, thus influencing directly the packet size and consequently the input rate to the source output buffer, whose objective is to match the *NNRC* calculated sending rate. In this respect, the source output buffer is prevented from being congested and/or starved.

We have to stress that even though the *NNRC* influences directly latency through RTT regulation, it has no control on jitter, which is very important in the perceived user QoS. Jitter may be controlled by incorporating a jitter buffer placed at the user endpoint to smooth out any variations in its size. The *NNRC* framework sees the jitter buffer as an additional buffer in the path, operating towards avoiding congestion collapse and throughput from starvation.

11.4   Illustrative Example

To evaluate the performance of a *NNRC* source enhanced with application adaptation, the design methodology developed herein is illustrated via simulations performed on the multi-bottleneck link network configuration described in Chapter 10 and is schematically presented in Figure 10.1.

11.4.1　Application Adaptation Implementation Details

The user satisfaction function $F(y)$ was taken as

$$F(y_1, y_2) = H(y_1, y_2) + R(y_1, y_2) \tag{11.12}$$

where

$$H(y_1, y_2) = \frac{1}{1 + e^{-10y_1 + 6.0}} + \frac{1}{1 + e^{-10y_2 + 6.0}}$$

$$R(y_1, y_2) = \frac{5}{1 + e^{-10y_1 + 6.0}} \frac{5}{1 + e^{-10y_2 + 6.0}}.$$

The selection of $F(y_1, y_2)$ as in (11.12) obviously satisfies (11.11). Moreover, the term H relates the user satisfaction individually to every media characteristic, whereas R models interdependencies between media characteristics and user satisfaction. The fact that (11.12) comprises a logical selection becomes evident by plotting $F(y_1, y_2)$. The regions in Figure 11.2 indicated

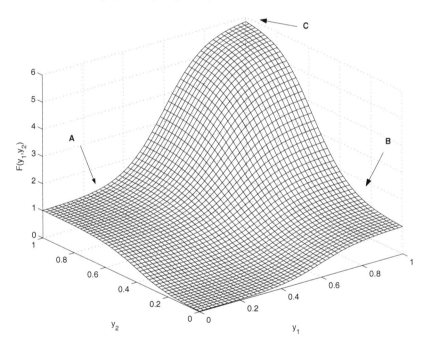

FIGURE 11.2
Three-dimensional plot of the satisfaction function (11.12).

by A, B are mainly produced by the term H and express qualitative dependencies of the form: if y_1 is small then the effect of y_2 in user satisfaction is reduced. The region indicated by C expresses the behavior of $F(y_1, y_2)$ when y_1 and y_2 are of sufficient content.

We have considered a bandwidth constraint $N(y_1, y_2)$ of the form:

$$N(y_1, y_2) = 80y_1y_2. \tag{11.13}$$

Assuming a video application where y_1 denotes the frame size and y_2 the bits per pixel, (11.13) is a logical selection.

To produce the required values of the media characteristics y_1, y_2 the application QoS controller is implemented with $a = 1$ and $S_i(y) = [S(y_i)\ S(y_i)^2\ S(y_i)^3]$. The sigmoid functions $S(y_i)$ were chosen to be

$$S(y_i) = \frac{1}{1 + e^{-y_i + 0.5}} > 0, \quad i = 1, 2.$$

For the user satisfaction level, it is assumed that $F_{min} = 0.75$, reflecting a user that can be satisfied if she/he receives the application with quality at 75% of the maximum attainable.

11.4.2   Simulation Study

We have materialized a scenario focusing on the effects caused by bursty traffic. More specifically, for the network configuration depicted in Figure 10.1 with link bandwidths 1Gps and maximum link buffer capacities 3000 packets, we consider altering the average rate of the CBR source in the interval [100, 550]Mps thus reducing the available link bandwidth from 900Mps to 450Mps. Furthermore, a 10% variation on the average rate of the CBR source is also allowed.

The available bandwidth $N_{max}(t)$ as provided by the *NNRC* framework is demonstrated in Figure 11.3.

The simulation results are presented in Figures 11.4 and 11.5, where critical parameters of the problem are plotted versus time. Figure 11.4 demonstrates the media characteristics y_1 and y_2, produced by the QoS control. The satisfaction error $(F(y_1, y_2) - F_{min})$ and the bandwidth error $(N(y_1, y_2) - N_{max}(t))$ are depicted in Figures 11.5.a and 11.5.b, respectively. It is clear that the system keeps the user satisfied, while not violating the bandwidth constraint $N_{max}(t)$ which varies significantly and abruptly.

11.5   Concluding Comments

Meeting user satisfaction in multimedia applications is a challenging as well as an important problem. Its complexity becomes evident if we formulated it as a nonlinear multivariable optimization problem with time-varying constraints.

In this chapter the problem is attacked by combining the time-sensitive congestion control framework *NNRC*, that directly influences latency through regulating the per packet RTT, with an application adaptation methodology

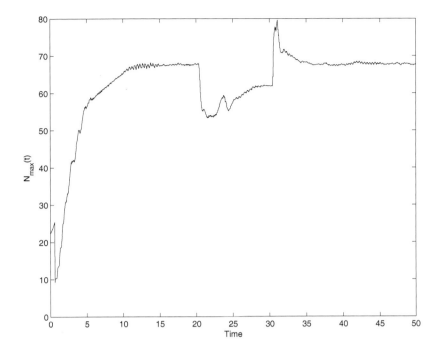

FIGURE 11.3
The available bandwidth $N_{max}(t)$ as produced by the *NNRC* framework.

that, based on dynamic neural networks and adaptive control techniques, provides those media characteristic values that enforce user satisfaction without violating the network induced constraints.

Simulation studies on a network topology of sufficient complexity highlight the basic ingredients of the proposed methodology.

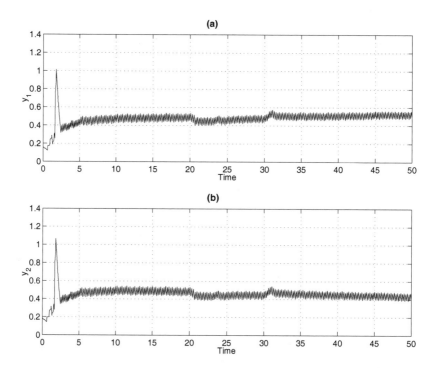

FIGURE 11.4
Simulation results: a) media characteristic y_1; b) media characteristic y_2.

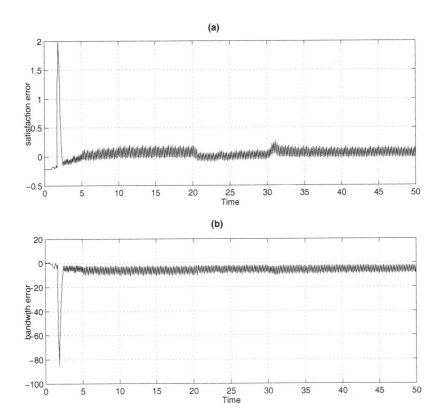

FIGURE 11.5
Simulation results: a) user satisfaction error $e = F(y_1, y_2) - F_{min}$ and b) bandwidth error $N(y_1, y_2) - N_{max}(t)$.

Part III

Appendices

A

Dynamic Systems and Stability

CONTENTS

This appendix presents necessary material related to the analysis and design of robust adaptive packet transmission rate controllers. It is therefore important for the unfamiliar reader to thoroughly review all the topics included to obtain the background required to fully understand Chapters 7 - 10. Further details may also be found in [11], [32], [94], [137], [161], [273], [289].

A.1 Vectors and Matrices

We start by providing the norm definition.

Definition A.1 *If $x \in \Re^n$, the vector norm of x is a real-valued function equipped with the following properties:*

- $|x| \geq 0$ *with* $|x| = 0$ *if and only if* $x = 0$

- $|ax| = |a||x|$, $\forall a \in \Re$ *and* $\forall x \in \Re^n$

- $|x + y| \leq |x| + |y|$, $\forall x, y \in \Re^n$ *(triangle inequality).*

Norms measure the size of elements in a set. Similarly, $|x - y|$ denotes the distance between the vectors x and y.

Definition A.2 *Let* $|\cdot|$ *be a vector norm. Then for each matrix* $A \in \Re^{m \times n}$ *the quantity* $\|A\|$ *defined by:*

$$\|A\| = \sup_{x \in \Re^n - \{0\}} \frac{|Ax|}{|x|} = \sup_{|x| \leq 1} |Ax| = \sup_{|x| = 1} |Ax|$$

is called the induced (matrix) norm of A.

The induced matrix norm satisfies the properties of Definition A.1. In addition the following properties of the induced norm will be frequently used in the text:

- $|Ax| \leq \| A \| \, |x|, \; \forall x \in \Re^n$

- $\|A + B\| \leq \| A \| + \| B \|$

- $\| AB \| \leq \| A \| \| B \|$

where A,B are arbitrary matrices of compatible dimensions. Some of the most commonly used vector and matrix norms are summarized below:

- $|x|_\infty = \max |x_i|$

- $|x|_1 = \sum_i |x_i|$

- $|x|_2 = \left(\sum_i |x_i|^2 \right)^{1/2}$

- $\|A\|_\infty = \max_i \sum_j |a_{ij}|$

- $\|A\|_1 = \max_j \sum_i |a_{ij}|$

- $\|A\|_2 = \left(\lambda_{max}(A^T A) \right)^{1/2}$

where x_i and a_{ij} denote the $i - th$ element of $x \in \Re^n$ and the $ij - th$ element of the $A \in \Re^{m \times n}$ respectively, while with $\lambda_{max}(C)$, $C \in \Re^{m \times n}$ we denote the maximum eigenvalue of the C matrix.

Generally, given a vector $x \in \Re^n$ with elements x_i, $i = 1, 2, ..., n$ we define the $p - norm$ of x as

$$|x|_p = \left(\sum_{i=1}^{n} |x_i|^p \right)^{1/p}, \quad p \in [1, \infty).$$

Example A.1 *Let* $x = [1 \; 2 \; 0 \; 3]^T$. *Then we have:*

$$|x|_\infty = 3, \quad |x|_1 = 6, \quad |x|_2 = \sqrt{14}.$$

If

$$A = \begin{bmatrix} 1 & 0 \\ 1 & 2 \\ 0 & 1 \end{bmatrix}, \quad B = \begin{bmatrix} 0 & 1 \\ 2 & 3 \end{bmatrix}$$

then we obtain:

$$\begin{array}{llll}
\|A\|_\infty = 3, & \|A\|_1 = 3, & \|A\|_2 = \sqrt{6} \\
\|B\|_\infty = 5, & \|B\|_1 = 4, & \|B\|_2 = \sqrt{13.708} \\
\|AB\|_\infty = 11, & \|AB\|_1 = 11, & \|AB\|_2 = \sqrt{78.695}.
\end{array}$$

In this manuscript, whenever a vector or a matrix norm is used without an explicit specification, the $2 - norm$ (Euclidean norm) will be implied.

It is to be noted that all vector norms are equivalent in that there exist constants $a, b \in \Re$ such that

$$a|x|_q \leq |x|_p \leq b|x|_q, \quad \forall p, q, \in [1, \infty].$$

For instance

$$\begin{array}{ll}
|x|_\infty \leq |x|_2 \leq \sqrt{n}|x|_\infty & , \quad \forall x \in \Re^n \\
|x|_\infty \leq |x|_1 \leq n|x|_\infty & , \quad \forall x \in \Re^n.
\end{array}$$

Furthermore, given a matrix $A \in \Re^{m \times n}$ with elements a_{ij} then

$$\max_{i,j} |a_{ij}| \leq \|A\|_2 \leq \sqrt{mn} \max_{i,j} |a_{ij}|.$$

Example A.2 *Let*

$$A = \begin{bmatrix} 2 & 1 \\ 0 & 3 \end{bmatrix}.$$

Then $\max_{i,j} |a_{ij}| = 3$ *and* $\|A\|_2 = 3.26$. *Hence,*

$$3 = \max_{i,j} |a_{ij}| \leq \|A\|_2 \leq \sqrt{mn} \max_{i,j} |a_{ij}| = 6.$$

A.1.1 Positive Definite Matrices

Definition A.3 *A real symmetric matrix* $A \in \Re^{n \times n}$ *is called positive semidefinite (denoted $A \geq 0$) if*

$$x^T A x \geq 0, \quad \forall x \in \Re^n$$

and positive definite (denoted $A > 0$) if

$$x^T A x > 0, \quad \forall x \in \Re^n - \{0\}.$$

Definition A.4 *A symmetric matrix* $A \in \Re^{n \times n}$ *is called negative semidefinite (negative definite) if $-A$ is positive semidefinite (positive definite).*

We shall write $A > B$ or $A \geq B$ if $A - B$ is positive definite or positive semidefinite, respectively. A symmetric matrix $A \in \Re^{n \times n}$ is positive definite if and only if one of the following holds:

- All eigenvalues of A are positive (i.e., $\lambda_i(A) > 0, \;\; i = 1, 2, ..., n$).

- There exists a nonsingular matrix A_1 such that $A = A_1 A_1^T$.

- All leading principal submatrices of A have positive determinants.

- For all $x \in \Re^n$, $\exists a > 0 : x^T A x \geq a|x|^2$.

If $A > 0$ then A^{-1} exists and satisfies $A^{-1} > 0$. If $A, B \in \Re^{n \times n}$ are positive semidefinite (definite) matrices, then the matrix $\kappa A + \lambda B$ is also positive semidefinite (definite), $\forall \kappa, \lambda > 0$. If $A \in \Re^{n \times n}$ is positive semidefinite and $C \in \Re^{m \times n}$ has $rank(C) = m$, then $CAC^T > 0$. Every positive definite matrix A can be written as $A = CC^T$ where C is square and invertible.

Positive definite matrices can be used to define the vector norm:

$$|x|_A^2 = x^T A x, \;\; A > 0.$$

In this direction notice that $|x|_A \geq 0$ and $|x|_A = 0$ if and only if $x = 0$. Furthermore, $|ax|_A = |a||x|_A, \; \forall a \in \Re$. Hence the first two properties of a vector norm (see Definition A.1) are trivially satisfied. We shall show that $|x + y|_A \leq |x|_A + |y|_A, \; \forall x, y \in \Re^n$ and $A > 0$. Notice, however that

$$\begin{aligned}
|x + y|_A &= \left((x + y)^T A (x + y)\right)^{1/2} \\
&= \left(|x^T A x + 2x^T A y + y^T A y|\right)^{1/2}.
\end{aligned}$$

Moreover,

$$\begin{aligned}
x^T A y &= x^T (A^{1/2})^T A^{1/2} y \\
&= (A^{1/2} x)^T (A^{1/2} y).
\end{aligned}$$

In addition

$$|x^T A y| \leq (x^T A x)^{1/2} (y^T A y)^{1/2}.$$

Hence,

$$\begin{aligned}
|x + y|_A &\leq \left(\left((x^T A x)^{1/2} + (y^T A y)^{1/2}\right)^2\right)^{1/2} \\
&= |x|_A + |y|_A.
\end{aligned}$$

If $A \geq 0$ then

$$\lambda_{min}(A)|x|^2 \leq x^T A x \leq \lambda_{max}(A)|x|^2, \;\; \forall x \in \Re^n$$

with $\lambda_{min}(A)$, $\lambda_{max}(A)$ being the minimum and maximum eigenvalues of A, respectively. Further, if $A \geq 0$ then

$$\|A\|_2 = \lambda_{max}(A)$$

and if $A > 0$ then

$$\|A^{-1}\|_2 = \frac{1}{\lambda_{min}(A)}.$$

It should be stressed that if $A > 0$ and $B \geq 0$, then $A + B > 0$. However, it is not generally correct that $AB \geq 0$. If $A \in \mathcal{R}_n$ with element a_{ij} and $A > 0$ then

$$\|A\|_2 \leq tr\{A\} \leq n\|A\|_2$$

where the trace operator defined by

$$tr\{A\} = \sum_{i=1}^{n} a_{ii}$$

has the properties:

- $tr\{AB\} = tr\{BA\}$,
- $tr\{A + B\} = tr\{A\} + tr\{B\}$, $\forall A, B \in \mathcal{R}^{n \times n}$,
- $tr\{yx^T\} = x^T y$, $\forall x, y \in \mathcal{R}^{n \times 1}$.

A.2 Signals

For a signal (i.e., a function of time), norms may also be used to define its magnitude. Given a continuous, vector-valued function $x(t) : [0, +\infty) \to \mathcal{R}^n$ its $\mathcal{L}_p - norm$ is calculated as:

$$\|x(t)\|_p = \left(\int_0^\infty |x(t)|^p dt \right)^{1/p}, \quad p \in [1, +\infty).$$

We shall say that $x \in \mathcal{L}_p$ when $\|x(t)\|_p$ exists, in other words when $\|x(t)\|_p$ is finite. The $\mathcal{L}_\infty$ norm is defined as:

$$\|x(t)\|_\infty = \sup_{t \geq 0} |x(t)|.$$

As in the $\mathcal{L}_p$ norm case, we shall say that $x \in \mathcal{L}_\infty$ when $\|x(t)\|_\infty$ exists.

Example A.3 *Consider* $x(t) = sin(t)$, $t \geq 0$. *It is true that* $x \in \mathcal{L}_\infty$. *However,* $x \notin \mathcal{L}_2$. *It is not difficult to verify also that* $e^{-t} \in \mathcal{L}_2$ *and that* $e^t \notin \mathcal{L}_\infty$.

Similarly with the continuous time case, for sequences the ℓ_p norm is defined as:

$$\|x\|_p = \left(\sum_{i=1}^{\infty} |x(i)|^p \right)^{1/p}, \quad p \in [1, \infty)$$

and

$$\|x\|_\infty = \sup_{i \geq 1} |x(i)|$$

where $x(i) \in \Re$ denotes the $i-th$ element of the sequence. We shall say $x \in \ell_p$ $(x \in \ell_\infty)$ if $\|x\|_p$ $(\|x\|_\infty)$ exists.

If for a signal $x(t) \in \mathcal{L}_1 \cap \mathcal{L}_\infty$ then $x(t) \in \mathcal{L}_p$, $\forall p \in [1, \infty]$. The following properties will be proven useful.

1. **Hölder's Inequality:** If $x(t) \in \mathcal{L}_p$ and $y(t) \in \mathcal{L}_q$, $\forall p, q \in [1, \infty]$ with $\frac{1}{p} + \frac{1}{q} = 1$, then $x(t)y(t) \in \mathcal{L}_1$ and

$$\|x(t)y(t)\|_1 \leq \|x(t)\|_p \|y(t)\|_q.$$

   In the special case $p = q = 2$, Hölder's inequality reduces to the Schwartz inequality i.e.,

$$\|x(t)y(t)\|_1 \leq \|x(t)\|_2 \|y(t)\|_2.$$

2. **Minkowski Inequality:** If $x(t), y(t) \in \mathcal{L}_p$ for $p \in [1, \infty]$ then $x(t) + y(t) \in \mathcal{L}_p$ and

$$\|x(t) + y(t)\|_1 \leq \|x(t)\|_p + \|y(t)\|_p.$$

3. **Young's Inequality:** For every $x(t), y(t) \in \Re$ the following is true:

$$x(t)y(t) \leq \frac{1}{\gamma} x^2(t) + \gamma y^2(t), \quad \forall t \geq 0, \ \forall \gamma > 0.$$

Example A.4 *Given $x(t), y(t) : \Re^+ \to \Re^n$, then if $x(t) \in \mathcal{L}_p$ and $y(t) \in \mathcal{L}_\infty$ for some $p \in [0, \infty)$, then $y^T(t)x(t) \in \mathcal{L}_p$.*

Proof: As $y(t) \in \mathcal{L}_\infty$, there exists a finite $k > 0$ such that $\sup_{t \geq 0} |y(t)| \leq k$. To continue, it is true that

$$
\begin{aligned}
\|y^T(t)x(t)\|_p &= \left(\int_0^\infty |y^T(\tau)x(\tau)|^p d\tau \right)^{1/p} \\
&\leq \left(\int_0^\infty |y^T(\tau)|^p |x(\tau)|^p d\tau \right)^{1/p} \\
&\leq \left(\int_0^\infty k^p |x(\tau)|^p d\tau \right)^{1/p} \\
&= k\|x(t)\|_p.
\end{aligned}
$$

Hence, $y^T(t)x(t) \in \mathcal{L}_p$.

A.3   Functions

In this section we shall review some basic function properties borrowed from real analysis.

A.3.1   Continuity

Definition A.5 *(Continuity) A function* $f : [0, \infty) \to \Re$ *is continuous on* $[0, \infty)$ *if for any given* $\varepsilon > 0$ *there exists a* $\delta(\varepsilon, t_o) > 0$ *such that* $\forall t_o, t \in [0, \infty)$ *satisfying* $|t - t_o| < \delta(\varepsilon, t_o)$ *then* $|f(t) - f(t_o)| < \varepsilon$.

Example A.5 *The function* $f(t) = cos(t)$ *is continuous* $\forall t \geq 0$. *However, the unit step function:*

$$f(t) = \begin{cases} 1, & t \geq 0 \\ \\ 0, & t < 0 \end{cases}$$

is discontinuous at $t = 0$ *according to Definition A.5, taking* $t_o = 0$ *we obtain* $|f(t) - 0| = 1, \quad \forall t \geq 0$. *Therefore, the definition does not hold* $\forall \varepsilon > 0$ *(e.g., for* $\varepsilon < 1$*).*

Definition A.6 *(Uniform Continuity) A function* $f : [0, \infty) \to \Re$ *is uniformly continuous on* $[0, \infty)$, *if for any given* $\varepsilon > 0$ *there exists a* $\delta(\varepsilon) > 0$ *such that* $\forall t_o, t \in [0, \infty)$ *satisfying* $|t - t_o| < \delta(\varepsilon)$ *then* $|f(t) - f(t_o)| < \varepsilon$.

Obviously, the difference between continuity and uniform continuity is that in the latter, δ is independent of t_o. As an example, the function $f(t) = \frac{1}{t}$ is continuous on $(0, \infty)$ but not uniformly continuous on the same interval, as no constant δ independent of t_o can be found satisfying $|f(t) - f(t_o)| < \varepsilon, \quad \forall \varepsilon > 0$ for an arbitrarily close to zero t_o.

An easy way of verifying uniform continuity is to check the boundedness of its $\frac{df(t)}{dt}$. Using this property it is straightforward to verify that the unit step is a discontinuous function.

Further, note that a uniformly continuous function is also continuous.

Definition A.7 *(Piecewise Continuity) A function* $f : [0, \infty) \to \Re$ *is piecewise continuous on* $[0, \infty)$, *if it is continuous on any finite interval* $[t_i, t_{i+1}] \subset [0, \infty)$ *except for a finite number of points.*

Notice that the unit step function is not continuous, but it is piecewise continuous.

Definition A.8 *(Absolute Continuity) A function* $f : [a, b] \to \Re$ *is absolutely continuous if and only if, for any* $\varepsilon > 0$ *there is a* $\delta > 0$ *such that*

$$\sum_{i=1}^{n} |f(a_i) - f(b_i)| < \varepsilon$$

for any finite collection of subintervals $(a_i, b_i) \subset [a, b]$ *with* $\sum_{i=1}^{n} |a_i - b_i| < \delta$.

Definition A.9 *(Lipschitz Continuity) A function* $f : [a, b] \to \Re$ *is said to be Lipschitz continuous or simply Lipschitz on* $[a, b]$ *if there exists a constant* $k \geq 0$ *(called the Lipschitz constant) such that* $|f(x_1) - f(x_2)| \leq k|x_1 - x_2|$, $\forall x_1, x_2 \in [a, b]$.

Clearly, all Lipschitz functions have a maximum slope on every point of their domain equal to the Lipschitz constant. If f is Lipschitz on $[a, b]$, then it is absolutely continuous. Moreover, consider a Lipschitz function f on $[a, b]$, with Lipschitz constant $k > 0$. Then given any $\varepsilon > 0$ we may choose $\delta = \varepsilon/k > 0$. If $x_1, x_2 \in [a, b]$ such that $|x_1 - x_2| < \delta$, then $|f(x_1) - f(x_2)| < k\delta = k\left(\frac{\varepsilon}{k}\right) = \varepsilon$. Therefore, according to the Definition A.6, f is uniformly continuous on $[a, b]$.

A.3.2 Differentiation

A function $f : \Re^n \to \Re^m$ is continuously differentiable, if the first partial derivatives of $f(x)$ with respect to x are continuous functions of x. Given $f : \Re^n \to \Re^m$, its Jacobian matrix is defined as:

$$\frac{\partial f}{\partial x} = \begin{bmatrix} \frac{\partial f_1}{\partial x_1} & \frac{\partial f_1}{\partial x_2} & \cdots & \frac{\partial f_1}{\partial x_n} \\ \frac{\partial f_2}{\partial x_1} & \frac{\partial f_2}{\partial x_2} & \cdots & \frac{\partial f_2}{\partial x_n} \\ \vdots & \vdots & \ddots & \vdots \\ \frac{\partial f_m}{\partial x_1} & \frac{\partial f_m}{\partial x_2} & \cdots & \frac{\partial f_m}{\partial x_n} \end{bmatrix}.$$

For a scalar function $f(x, y) : \Re^n \times \Re^m \to \Re$ its gradient with respect to x is defined as:

$$\nabla_x f(x, y) = \begin{bmatrix} \frac{\partial f}{\partial x_1} & \frac{\partial f}{\partial x_2} & \cdots & \frac{\partial f}{\partial x_n} \end{bmatrix}.$$

If f is only a function of $x \in \Re^n$ then we shall use the notation $\nabla f(x)$ to denote its gradient.

If $f : \Re^n \to \Re^m$ is differentiable for each point in $\mathcal{A} \subseteq \Re^n$, where $\mathcal{A}$ is a convex[1] set, then $\forall x_1, x_2 \in \mathcal{A}$ there exists some ξ that belongs to the line segment connecting x_1 with x_2 such that:

$$f(x_2) - f(x_1) = \left.\frac{\partial f(x)}{\partial x}\right|_{x=\xi} (x_2 - x_1).$$

This is the well-known mean value theorem.

---

[1] $\mathcal{A}$ is a convex set if $\forall x_1, x_2 \in \mathcal{A}$ and $\lambda \in [0, 1]$, we have $\lambda x_1 + (1 - \lambda)x_2 \in \mathcal{A}$.

A.3.3 Convergence

A function $f : [0, \infty) \to \Re$ that is bounded from below and is nonincreasing, or is bounded from above and is nondecreasing, has a limit as $t \to \infty$. Notice that we may not conclude that $f(t) \in \mathcal{L}_\infty$. Consider for example the function $f(t) = 1/t$ for which $\lim_{t\to\infty} f(t) = 0$ with $f(0) = \infty$. However, if additionally $f(0)$ is finite, then $f(t) \in \mathcal{L}_\infty$ since $f(t) \leq f(0)$, $\forall t \geq 0$ (the nonincreasing case) and $f(t)$ is bounded from below.

The following statements are useful in the understanding and analysis of the behavior of adaptive systems.

Statement A.1 $\lim_{t\to\infty} \dot{f}(t) = 0$ *does not imply that* $f(t)$ *has a limit as* $t \to \infty$.

Example A.6 *Consider the function* $f(t) = \sin(\sqrt{t})$, $t \geq 0$, *for which*

$$\dot{f}(t) = \frac{\cos(\sqrt{t})}{2\sqrt{t}}.$$

Clearly $\lim_{t\to\infty} \dot{f}(t) = 0$. *However,* $f(t)$ *has no limit as* $t \to \infty$.

Therefore, if for a function $f(t)$ it holds $\lim_{t\to\infty} \dot{f}(t) = 0$ we can only conclude that the rate of change of $f(t)$ keeps reducing with t.

Statement A.2 $\lim_{t\to\infty} f(t) = k$ *for some constant* $k \in \Re$ *does not imply that* $\lim_{t\to\infty} \dot{f}(t) = 0$.

Example A.7 *Consider the function*

$$f(t) = \frac{\sin(t^n)}{t}, \quad n \geq 2, \quad t \geq 0.$$

Obviously

$$\lim_{t\to\infty} f(t) = 0.$$

However,

$$\dot{f}(t) = n t^{n-2} \cos t^n - \frac{1}{t^2} \sin(t^n), \quad \forall t \geq 0$$

which has no limit for $n \geq 2$ *and becomes unbounded for* $n > 2$.

The next lemma is useful in the development of the control theoretic results of this book.

Lemma A.1 *Let* $\eta, V : [0, \infty) \to \Re$. *Then*

$$\dot{V} \leq -\gamma V + \eta, \quad \forall t \geq 0$$

with γ *a positive constant, implies that*

$$V(t) \leq e^{-\gamma t} V(0) + \int_0^t e^{-\gamma(t-\tau)} \eta(\tau) d\tau, \quad \forall t \geq 0.$$

The following lemma and its corollary are central in proving convergence results.

Lemma A.2 *(Barbălat) Let $\eta(t)$ be a uniformly continuous function. If $\lim\limits_{t\to\infty} \int_0^t \eta(\tau)d\tau$ exists and is finite, then $\lim\limits_{t\to\infty} \eta(t) = 0$.*

Corollary A.1 *If $\eta, \dot{\eta} \in \mathcal{L}_\infty$ and $\eta \in \mathcal{L}_p$ for some $p \in [1, \infty)$, then $\lim\limits_{t\to\infty} f(t) = 0$.*

A.3.4 Function Properties

Definition A.10 *A continuous function $f : [0, r] \to \Re^+$, (or on $[0, \infty)$) is said to belong to class $\mathcal{K}$ and is denoted by $f \in \mathcal{K}$ if*

- $f(0) = 0$

- f is strictly increasing on $[0, r]$, (or on $[0, \infty)$).

Definition A.11 *A continuous function $f : [0, \infty] \to \Re^+$ is said to belong to class $\mathcal{K}_\infty$ (denoted by $f \in \mathcal{K}_\infty$) if*

- $f \in \mathcal{K}$

- $\lim\limits_{t\to\infty} f(t) = \infty$.

Definition A.12 *A continuous function $f : \mathcal{A} \times \Re^+ \to \Re^+$ is said to belong to class $\mathcal{KL}$, if $f(x, y) \in \mathcal{K}$ for each fixed y and $f(x, y)$ is decreasing with respect to y for each fixed x with $\lim\limits_{x\to\infty} f(x, y) = 0$.*

Example A.8 *The function $f_1(x) = x^2$ is strictly increasing on $[0, \infty)$, $f_1(0) = 0$ and $\lim\limits_{x\to\infty} f_1(x) = \infty$. Therefore, $f_1 \in \mathcal{K}$ and $f_1 \in \mathcal{K}_\infty$. However, the function $f_2(x) = \frac{x^2}{1+x^2}$ is strictly increasing, $f_2(0) = 0$ but $\lim\limits_{x\to\infty} f_2(x) = 1 \neq \infty$. Hence $f_2 \in \mathcal{K}$, $f_2 \notin \mathcal{K}_\infty$.*

Definition A.13 *Two functions f_1, f_2 of class $\mathcal{K}$ defined on $[0, r]$ (or on $[0, \infty)$), are said to be of the same order, if there exist positive constants $c_1, c_2 > 0$ such that:*

$$c_1 f_1(x) \leq f_2(x) \leq c_2 f_1(x), \quad \forall x \in [0, r], \text{ (or on } [0, \infty)).$$

Definition A.14 *A function $V(t, x) : \Re^+ \times \mathcal{B}_r \to \Re$, where $\mathcal{B}_r = \{x \in \Re^n : |x| < r\}$, with $V(t, 0) = 0$, $\forall t \geq 0$ is positive definite, if there exists a continuous function $\gamma \in \mathcal{K}$ such that*

$$V(t, x) \geq \gamma(|x|), \forall t \geq 0, \forall x \in \mathcal{B}_r, r > 0.$$

Definition A.15 *A function $V(t, x) : \Re^+ \times \mathcal{B}_r \to \Re$, with $V(t, 0) = 0$, $\forall t \geq 0$ is said to be negative definite if $-V(t, x)$ is positive definite.*

The set $\mathcal{B}_r$ defined in Definition A.13 constitutes a ball of radius $r > 0$, centered at the origin $(x = 0)$.

Example A.9 *The function $V(t, x) = \frac{x^2}{1+x^2}$ is positive definite $\forall x \in \Re$, since $V(t, x) \geq \frac{x^2}{1+x^2} \in \mathcal{K}$. However, the function $V(t, x) = \frac{1}{t}x^2$ is not, as it is impossible to find a class $\mathcal{K}$ function to satisfy the condition of Definition A.14, $\forall t \geq 0$.*

Definition A.16 *A function $V(t, x) : \Re^+ \times \mathcal{B}_r \to \Re$ with $V(t, 0) = 0$, $\forall t \geq 0$ is said to be positive semidefinite, if $V(t, x) \geq 0$ for all $(t, x) \in \Re^+ \times \mathcal{B}_r$ for some $r > 0$. The function $V(t, x)$ will be called negative semidefinite if $V(t, x) \leq 0$.*

Definition A.17 *A function $V(t, x) : \Re^+ \times \mathcal{B}_r \to \Re$ with $V(t, 0) = 0$, $\forall t \geq 0$ is said to be decrescent, if there exists a function γ of class $\mathcal{K}$ such that $|V(t, x)| \leq \gamma(|x|)$, $\forall (t, x) \in \Re^+ \times \mathcal{B}_r$, for some $r > 0$.*

Example A.10 *The function $V(t, x) = \frac{1}{t}x^2$ is decrescent since $V(t, x) \leq |x|^2$, $\forall t \geq 0$ and $\forall x \in \Re$, whereas $V(t, x) = tx^2$, $\forall t \geq 0$ is not.*

Definition A.18 *A function $V(t, x) : \Re^+ \times \Re^n \to \Re$ with $V(t, 0) = 0$, $\forall t \geq 0$ is called radially unbounded, if there exists a class $\mathcal{K}_\infty$ function γ such that $V(t, x) \geq \gamma(|x|)$, $\forall x \in \Re^n$ and $\forall t \geq 0$.*

Example A.11 *The function $V(t, x) = \frac{x^2}{1+x^2}$ is not radially unbounded because $V(t, x) \leq 1$, $\forall x \in \Re$. Hence, it is impossible to find a $\gamma \in \mathcal{K}_\infty$ satisfying $V(t, x) \geq \gamma(|x|)$, $\forall x \in \Re$. However, for the function $V(t, x) = (1 + t)x^2$ we can choose the $\mathcal{K}_\infty$ function $\gamma(|x|) = |x|^2$ which satisfies $V(t, x) \geq |x|^2$, $\forall x \in \Re$, $\forall t \geq 0$. Therefore, $V(t, x) = (1 + t)x^2$ is radially unbounded.*

Clearly, any radially unbounded function is also positive definite. However, the opposite is not necessarily true. Furthermore, from the Definition A.17 it can be easily concluded that any time independent function that is positive or negative definite is decrescent.

A.4 Dynamic Systems

We shall consider dynamic systems that can be described by a set of nonautonomous (time-varying) first order, nonlinear, differential equations of the form:

$$\dot{x} = f(t, x), \quad x(t_o) = x_o \tag{A.1}$$

where $x \in \Re^n$ and $f : [0, \infty) \times B_r \to \Re^n$. The parameter n is referred to as the system order. The vector x is called the state vector.

Assumption A.1 *For every* $x_o \in B_r$ *and every* $t_o \geq 0$, *the initial value problem (A.1) possesses a unique solution* $x(t; t_o, x_o)$.

The state trajectory in $\Re^n$ $x(t; t_o, x_o)$ is called a solution to (A.1) if

- $x(t; t_o, x_o) = x_o$,

- $\frac{dx(t;t_o,x_o)}{dt} = f(t, x(t; t_o, x_o))$.

In general, different t_o, x_o produce different solutions $x(t; t_o, x_o)$ of (A.1).

Definition A.19 *A point* $x^* \in \Re^n$ *is said to be an equilibrium point of (A.1) if*

$$f(t, x^*) = 0, \ \forall t \geq t_o \geq 0.$$

Definition A.20 *An equilibrium point* x^* *is called isolated, if there exists a ball centered at* x^* *of some radius* $r > 0$, $B_{x^*,r} = \{x \in \Re^n : |x - x^*| < r\}$ *that contains no equilibrium points other than* x^*.

Example A.12 *Consider the system defined by:*

$$\begin{aligned} \dot{x}_1 &= x_1 x_2 \\ \dot{x}_2 &= -x_2. \end{aligned}$$

It is obvious that $(x_1^*, x_2^*) = (0, 0)$ *is an equilibrium point. However, it is not isolated, since* $(x_1^*, x_2^*) = (a, 0)$, $\forall a \in \Re$ *is also an equilibrium point.*

In what follows we shall assume, with no loss of generality, that the equilibrium of interest x^* is isolated and that $x^* = 0$.

Example A.13 *Consider the system*

$$\dot{x} = (x - 1)^2, \ x \in \Re \tag{A.2}$$

for which $x^* = 1$ *is the equilibrium point. If we define* $\tilde{x} = x - x^* = x - 1$ *the original system equivalently becomes*

$$\dot{\tilde{x}} = \dot{x} = (\tilde{x} + 1 - 1)^2 = \tilde{x}^2 \tag{A.3}$$

which has $\tilde{x}^* = 0$ *as an equilibrium point. Therefore, instead of studying the behavior of (A.2) around* $x^* = 1$, *we can study, with no loss of generality, the behavior of (A.3) around* $\tilde{x}^* = 0$.

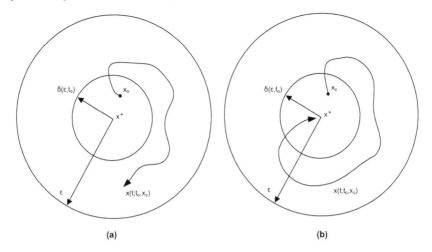

FIGURE A.1
Graphical representation of a) a stable equilibrium x^* and b) of an asymptotically stable equilibrium x^*.

A.4.1 Stability Definitions

We shall consider nonlinear, nonautonomous systems of the form (A.1).

Definition A.21 *The equilibrium point x^* of (A.1) is called stable, if $\forall t_o \geq 0$ and $\forall \varepsilon > 0$, $\exists \delta(\varepsilon, t_o) > 0$ such that*

$$|x_o - x^*| < \delta(\varepsilon, t_o) \Rightarrow |x(t; t_o, x_o) - x^*| < \varepsilon, \ \ \forall t \geq t_o \geq 0.$$

Definition A.22 *The equilibrium point x^* of (A.1) is called uniformly stable, if it is stable and δ in Definition A.21 is independent of t_o.*

Definition A.23 *The equilibrium point x^* of (A.1) is called asymptotically stable, if it is stable and $\exists \delta(t_o)$ such that*

$$|x_o - x^*| < \delta(t_o) \Rightarrow \lim_{t \to \infty} |x(t; t_o, x_o) - x^*| = 0.$$

Definition A.24 *The equilibrium point x^* of (A.1) is called uniformly asymptotically stable, if it is uniformly stable and $\forall \varepsilon > 0$ and any $t_o \geq 0$, $\exists \delta > 0$ (independent of t_o, ε) and a $T(\varepsilon) > 0$ such that:*

$$|x_o - x^*| < \delta \Rightarrow |x(t; t_o, x_o) - x^*| < \varepsilon, \ \forall t \geq t_o + T(\varepsilon).$$

Graphically, both the stability and asymptotic stability definitions are illustrated in Figure A.1.

Definition A.25 *The set of all initial conditions x_o for which $\lim_{t \to \infty} x(t; t_o, x_o) = x^*$, for some $t_o \geq 0$ is called the region of attraction.*

Definition A.26 *If $\exists \delta(t_o) > 0$ such that:*

$$|x_o - x^*| < \delta(t) \Rightarrow \lim_{t \to \infty} |x(t; t_o, x_o) - x^*| = 0$$

then the equilibrium x^* is said to be attractive.

Definition A.27 *The equilibrium point x^* of (A.1) is called uniformly asymptotically stable, if it is uniformly stable and $\forall \varepsilon > 0$, $\forall t_o \geq 0$ there exists a $\delta > 0$ (independent of t_o, ε) as well as a $T(\varepsilon) > 0$ such that*

$$|x_o - x^*| < \delta \Rightarrow |x(t; t_o, x_o) - x^*| < \varepsilon, \ \ \forall t \geq t_o + T(\varepsilon).$$

Definition A.28 *The equilibrium point x^* of (A.1) is exponentially stable if there exists a $\gamma > 0$ (called convergence rate) and $\forall \varepsilon > 0$ there exists a $\delta(\varepsilon) > 0$ such that*

$$|x_o - x^*| < \delta(\varepsilon) \Rightarrow |x(t; t_o, x_o) - x^*| \leq \varepsilon e^{-\gamma(t-t_o)}, \ \ \forall t \geq t_o.$$

Definition A.29 *The equilibrium point x^* of (A.1) is unstable if it is not stable.*

    In case (A.1) is autonomous (i.e., $\dot{x} = f(x)$) the definitions of stability/uniform stability and asymptotic stability/uniform asymptotic stability are equivalent. In a few words, the equilibrium is stable if starting close to it the system solution always stays close to it. Whereas it is asymptotically stable, if starting close to it, the solution converges to it. It is desirable to design control systems that do not exhibit instabilities. Stability is a property related to equilibrium points, not to a system. We shall refer to a system as being stable if all its equilibriums are stable.

Example A.14 *Let us consider the system $\dot{x} = -(x - 2)$, $x \in \Re$ for which $x^* = 2$ is the unique equilibrium point. Defining $\tilde{x} = x - 2$ the original system equivalently becomes*

$$\dot{\tilde{x}} = -\tilde{x}.$$

Notice that

$$\tilde{x}(t; t_o, \tilde{x}_o) = e^{-(t-t_o)}\tilde{x}_o, \ \ \forall t \geq t_o \geq 0.$$

Therefore, the solution $\tilde{x}(t; t_o, \tilde{x}_o)$ converges exponentially (with convergence rate 1) to the equilibrium point $\tilde{x}_o^* = 0$ for all $\tilde{x}_o \in \Re$. In this case the set $\Re$ defines the region of attraction.

A.4.2 Boundedness Definitions

The next four definitions hold under Assumption A.1.

Definition A.30 *The solution $x(t; t_o, x_o)$ of (A.1) is said to be bounded if there exists a $\gamma > 0$ (possibly dependent on the solution) such that:*

$$x(t; t_o, x_o) \leq \gamma, \quad \forall t \geq t_o \geq 0.$$

Definition A.31 *The solution $x(t; t_o, x_o)$ of (A.1) is said to be uniformly bounded if for any $\xi > 0$ and $t_o \geq 0$, there exists a $\gamma(\xi) > 0$ such that:*

$$|x_o| < \xi \Rightarrow |x(t; t_o, x_o)| \leq \gamma(\xi), \quad \forall t \geq t_o.$$

Definition A.32 *The solution $x(t; t_o, x_o)$ of (A.1) is said to be uniformly ultimately bounded if there exists a $M > 0$ and if for any $\xi > 0$ and $t_o \geq 0$ there exists a $T(\xi) > 0$ such that:*

$$|x_o| < \xi \Rightarrow |x(t; t_o, x_o)| \leq M, \quad \forall t \geq t_o + T(\xi).$$

Definition A.33 *System (A.1) is said to be Lagrange stable if $\forall t_o \geq 0$ and $x_o \in \mathcal{B}_r$, the solution $x(t; t_o, x_o)$ is bounded.*

Lagrange stability implies that all solutions $x(t; t_o, x_o)$ of (A.1) are bounded irrespectively of t_o, x_o. This is conceptually different from the notion of a stable equilibrium, since there might exist $t_o \geq 0$ and $x_o \in \mathcal{B}_r$ (close to an unstable equilibrium) resulting in $x(t; t_o, x_o) \to \infty$ as $t \to \infty$.

The difference between a bounded solution and a uniformly bounded solution is that the upper bound γ is independent of t_o in the latter case. Therefore, uniform boundedness implies boundedness and thus Lagrange stability. However, the opposite is not necessary correct. For autonomous (i.e., time independent) systems, boundedness automatically implies uniform boundedness. In fact, nonuniformity is a problem related to nonautonomous systems.

Translating the uniform ultimate boundedness definition, if we start from inside the $M - zone$ (i.e., $\xi < M$) then the solution $x(t; t_o, x_o)$ stays within $\forall t \geq t_o \geq 0$. If however we originate outside the $M - zone$ ($|x_o| > M$) then there exists a finite time $T(\xi)$ after which the solution $x(t; t_o, x_o)$ is confined strictly within the $M - zone$ (i.e., $|x(t; t_o, x_o)| \leq M, \ \forall t \geq t_o + T(\xi)$). The term uniform indicates that T is independent of t_o, while the term ultimate indicates that the property $|x(t; t_o, x_o)| \leq M$ holds after a time lapse T.

In Figure A.2 the qualitative differences between uniform boundedness and uniform ultimate boundedness are graphically illustrated.

Notice that a uniform ultimate boundedness solution is always bounded even prior to $t_o + T$, as Figure A.2(b) clearly demonstrates. In that respect, uniform ultimate boundedness is a stronger property compared to uniform boundedness.

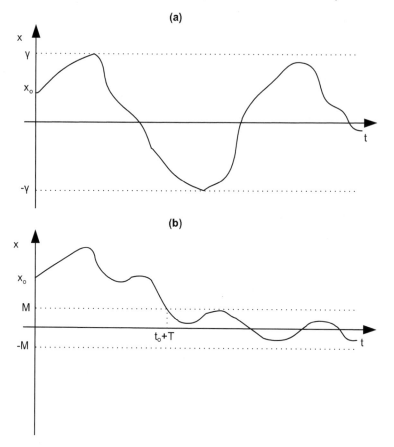

FIGURE A.2
Graphical illustration of the uniform boundedness (a) and the uniform ultimate boundedness (b) properties.

A.4.3   Stability Tools

The stability tools that follow are centered on A.M. Lyapunov's direct method
(also called Lyapunov's second method), which tries to answer questions regarding the stability of an equilibrium point of system (A.1), without residing
to the explicit knowledge of its solutions.

Theorem A.1 *(Lyapunov's Direct Method) Consider the nonlinear system
(A.1) which satisfies Assumption A.1. If there exists a positive function
$V(t,x) : [0, \infty) \times \mathcal{B}_r \to \Re_+$ for some $r > 0$, with continuous first order
partial derivatives and $V(t,0) = 0$, $\forall t \geq 0$, then the isolated equilibrium point
x^* of (A.1) is:*

- *Stable, if $\dot{V} \leq 0$.*

- *Uniformly stable, if V is decrescent and $\dot{V} \leq 0$.*

- *Uniformly asymptotically stable, if V is decrescent and $\dot{V} < 0$.*

- *Exponentially stable, if V is decrescent and there exist $\mathcal{K}$-functions, $\gamma_1, \gamma_2, \gamma_3$ of the same order such that:*

$$\gamma_1(|x|) \leq V(t, x) \leq \gamma_2(|x|)$$

$$\dot{V}(t, x) \leq -\gamma_3(|x|)$$

$\forall x \in \mathcal{B}_r$ *and* $t \geq 0$.

The statement on x^* being uniformly asymptotically stable can be equivalently rephrased as: there exist $\mathcal{K}$-functions $\gamma_1, \gamma_2, \gamma_3$ not of the same order such that

$$\gamma_1(|x|) \leq V(t, x) \leq \gamma_2(|x|)$$

$$\dot{V}(t, x) \leq -\gamma_3(|x|)$$

$\forall x \in \mathcal{B}_r$ and $t \geq 0$.

Theorem A.2 *Consider the nonlinear system (A.1) which satisfies Assumption A.1 for all $x_o \in \Re^n$. If there exists a function $V(t, x) : [0, \infty) \times \bar{\mathcal{B}}_r \to \Re_+$, where $\bar{\mathcal{B}}_r$ is the complement of $\mathcal{B}_r$ where $r > 0$ may be large, with continuous first order partial derivatives and if there exists $\gamma_1, \gamma_2 \in \mathcal{K}_\infty$ such that*

$$\gamma_1(|x|) \leq V(t, x) \leq \gamma_2(|x|)$$

$$\dot{V}(t, x) \leq 0(|x|)$$

$\forall |x| \geq r$ *and* $t \in [0, \infty)$, *then the solutions $x(t; t_o, x_o)$ of (A.1) are uniformly bounded. If in addition there exists a $\gamma_3 \in \mathcal{K}$ defined on $[0, \infty)$ and*

$$\dot{V}(t, x) \leq -\gamma_3(|x|), \ \forall |x| > r \text{ and } t \in [0, \infty)$$

then the solutions $x(t; t_o, x_o)$ are uniformly ultimately bounded with respect to the set $\{x \in \Re^n : |x| < r\}$.

Theorem A.3 *Consider the nonlinear system (A.1) which satisfies Assumption A.1 for all $x_o \in \Re^n$. If there exists a positive definite, decrescent and radially unbounded function $V(t, x) : [0, \infty) \times \Re^n \to \Re_+$ with continuous first order partial derivatives, then the equilibrium point x^* is:*

- *Globally uniformly asymptotically stable, if $\dot{V} < 0$.*

- *Globally exponentially stable, if there exist $\gamma_1, \gamma_2, \gamma_3 \in \mathcal{K}_\infty$ of the same order such that:*

$$\gamma_1(|x|) \leq V(t, x) \leq \gamma_2(|x|)$$

$$\dot{V}(t, x) \leq -\gamma_3(|x|).$$

The statement on x^* being globally uniformly asymptotically stable can be equivalently rephrased as: there exist $\mathcal{K}$-functions γ_1, γ_2 and a $\mathcal{K}_\infty$-functions γ_3 such that:

$$\gamma_1(|x|) \leq V(t,x) \leq \gamma_2(|x|)$$

$$\dot{V}(t,x) \leq -\gamma_3(|x|)$$

for all $x \in \Re^n$. Theorems A.1 and A.2 apply also to nonautonomous systems. However, in that case, the words "decrescent" and "uniform" can be dropped.

For autonomous systems

$$\dot{x} = f(x)$$

the following theorem applies.

Theorem A.4 *(LaSalle's Invariant Set Theorem) For autonomous systems satisfying Assumption A.1 for all $x \in \Re^n$. If there exists a positive definite and radially unbounded function $V(x) : \Re^n \to \Re_+$ with continuous first order derivatives satisfying:*

- $\dot{V}(x) \leq 0, \ \forall \in \Re^n,$

- *the origin $x = 0$ is the only invariant subset of $\{x \in \Re^n : \dot{V} = 0\},$*

then the equilibrium point $x^* = 0$ is asymptotically stable.

Recall that a set $\mathcal{A}$ is invariant in $\Re^n$ with respect to $\dot{x} = f(x)$, if every solution of $\dot{x} = f(x)$ starting in $\mathcal{A}$ remains in $\mathcal{A}$, for all t.

A function V that satisfies Theorem A.1-A.4 is called the Lyapunov function. The examples that follow clarify the use of the stability tools reported.

Example A.15 *Consider the system:*

$$\begin{aligned} \dot{x}_1 &= x_2 + cx_1 \\ \dot{x}_2 &= -x_1 + cx_2 \end{aligned} \tag{A.4}$$

where c is a constant. Notice that $x^* = (x_1^*, x_2^*) = (0,0)$ is the only equilibrium of the system. Choose the Lyapunov function candidate

$$V(x_1, x_2) = \frac{1}{2}x_1^2 + \frac{1}{2}x_2^2.$$

Notice that V is positive definite and radially unbounded. Its time derivative along the solution of (A.4) is

$$\dot{V} = c(x_1^2 + x_2^2).$$

If $c = 0$, then $\dot{V} = 0$ and therefore applying Theorem A.3 it is obtained that the equilibrium point $x^* = 0$ is globally uniformly stable.

If $c < 0$, then $\dot{V} < 0$. Hence, according to Theorem A.3 the equilibrium point $x^* = 0$ is globally uniformly asymptotically stable.

If $c > 0$, the equilibrium point $x^* = 0$ is unstable, because in this case V

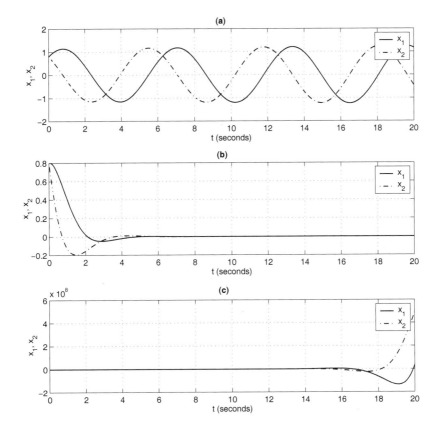

FIGURE A.3
State trajectories of (A.4) for different values of the parameter c. Both states
are initialized at 0.8. a) $c = 1$; b) $c = -1$; c) $c = 1$.

*is strictly increasing as $\dot{V} > 0$, $\forall t \geq 0$. Therefore the solutions of (A.4) will
go unbounded.*

*The aforementioned results are graphically illustrated in Figure A.3 which
plots the state trajectories of (A.4) for various values of the parameter c.*

Example A.16 *Consider the system:*

$$\dot{x} = -g(x), \quad g(0) = 0, \quad \forall x \in \Re \tag{A.5}$$

*where $xg(x) > 0$, $\forall x \neq 0$ and $x \in (-a, a)$. Notice that $x^* = 0$ is the only
equilibrium point. Let us now choose the Lyapunov function candidate:*

$$V(x) = \int_0^x g(y)dy.$$

For all $x \in (-a, a)$, $V(x)$ is continuously differentiable with $V(0) = 0$ and

$V(x) > 0, \forall x \neq 0$. *The time derivative of V along the solutions of (A.5) is*

$$\dot{V} = -g^2(x) < 0, \quad \forall x \in (-a, \ a) - \{0\}.$$

Therefore, according to Theorem A.1, the equilibrium point $x^* = 0$ is asymptotically stable.

     Take for example the system

$$\dot{x} = -x^3$$

where $g(x) = x^3$, which satisfies the stated conditions $xg(x) > 0, \forall x \neq 0$ and $x \in (-a, \ a)$. Also $g(0) = 0$. The above-mentioned analysis proves that the equilibrium point $x^* = 0$ is asymptotically stable. A graphical validation is presented in Figure A.4 which pictures the evolution of x with respect to time.

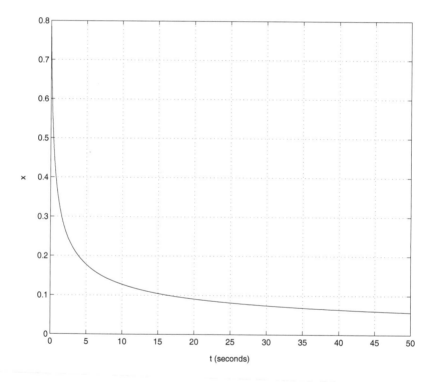

FIGURE A.4

Evolution of x with respect to time for (A.5) when $g(x) = x^3$. The state is initialized at $x(0) = 0.8$ and converges asymptotically to the equilibrium $x^* = 0$.

Example A.17 *Consider the system:*

$$\begin{aligned} \dot{x}_1 &= x_2 \\ \dot{x}_2 &= -a\sin x_1 - bx_2, \quad a, b > 0. \end{aligned} \tag{A.6}$$

To begin with, notice that $x^* = (x_1^*, x_2^*) = (0,0)$ is the only equilibrium point $\forall x_1 \in (-\frac{\pi}{2}, \frac{\pi}{2})$. Choose the Lyapunov function candidate

$$V(x) = a \int_0^x \sin z\, dz + \frac{1}{2} x^T P x$$

where P is a positive definite matrix and $x = [x_1\ x_2]^T$. Obviously,

$$
\begin{aligned}
x^T P x &= [x_1\ x_2] \begin{bmatrix} P_{11} & P_{12} \\ P_{21} & P_{22} \end{bmatrix} \begin{bmatrix} x_1 \\ x_2 \end{bmatrix} \\
&= P_{11} x_1^2 + P_{12} + P_{22} x_2^2.
\end{aligned}
$$

Therefore to satisfy the positive definiteness of P the following should hold:

$$P_{12} = P_{21}, \quad P_{11} > 0, \quad P_{11} P_{22} > P_{12}^2.$$

The time derivative of V along the solution of (A.6) is

$$
\begin{aligned}
\dot{V}(x) &= (P_{11} - P_{12}b) x_1 x_2 + a(1 - P_{22}) x_2 \sin x_1 \\
&= (P_{12} - P_{22}) x_2^2 - P_{12} a x_1 \sin x_1.
\end{aligned}
$$

In the aforementioned expression, notice that $x_2^2 > 0$, $\forall x_2 \in \Re$ and $x_1 \sin x_1 > 0$ for $x_1 > 0$. On the contrary, the terms $x_1 x_2$ and $x_2 \sin x_1$ are of indefinite sign. However, selecting

$$
\begin{aligned}
P_{11} &= b P_{12} \\
P_{22} &= 1
\end{aligned}
$$

the undesirable terms in $\dot{V}$ are deleted, resulting in

$$\dot{V}(x) = (P_{12} - b) x_2^2 - P_{12} a x_1 \sin x_1.$$

To guarantee $\dot{V}(x) < 0$ we should have $P_{12} < b$ and $P_{12} > 0$. Additionally, from the positive definiteness of P we also have $P_{11} P_{22} > P_{12}^2$. Thus, $P_{12}^2 - b P_{12} < 0$ or $0 < P_{12} < b$. Therefore, selecting $P_{12} = b/2$ we obtain

$$\dot{V}(x) < 0, \quad \forall x \in \Re^2 : |x_1| < \pi.$$

Hence, according to Theorem A.1 the equilibrium point $(x_1^*, x_2^*) = (0,0)$ is asymptotically stable.

In Figure A.5 we have simulated system (A.6) with $a = b = 1$. The system initializes at $x_1(0) = x_2(0) = 0.1$ and converges asymptotically to the equilibrium point $(x_1^*, x_2^*) = (0,0)$.

Example A.18 *Consider the system:*

$$
\begin{aligned}
\dot{x}_1 &= -x_1 + x_2 + x_1 x_2^3 \\
\dot{x}_2 &= -x_1 - x_1^2 x_2^2.
\end{aligned}
\tag{A.7}
$$

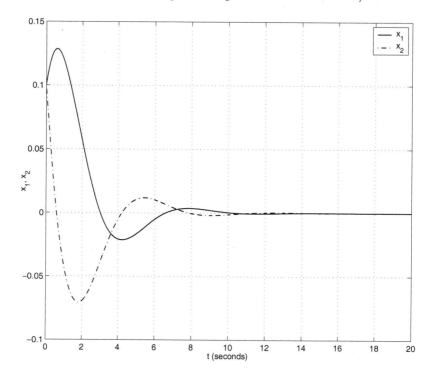

FIGURE A.5
State evolution of (A.6) with $a = b = 1$ with respect to time. Both states
initialize at 0.1 and converge asymptotically to the equilibrium $(x_1^*, x_2^*) =
(0,0)$.

*Notice that $(x_1^*, x_2^*) = (0,0)$ is the only equilibrium point of (A.7). Further-
more, choose the Lyapunov function candidate*

$$V(x_1, x_2) = \frac{1}{2}x_1^2 + \frac{1}{2}x_2^2.$$

Taking the time derivative of V along the solutions of (A.7) we obtain:

$$\dot{V} = -x_1^2 \leq 0.$$

*Hence, according to Theorem A.1, the equilibrium point $x^* = (0,0)$ is uni-
formly stable. However, if we define*

$$\mathcal{A} = \{x \in \Re^2 : x_1 = 0\}$$

*then on $\mathcal{A}$, $\dot{x}_1 = x_2$. Therefore, any solution starting from $\mathcal{A}$ with $x_2 \neq 0$
leaves $\mathcal{A}$. In other words, $(x_1, x_2) = (0,0)$ is the only invariant set of $\mathcal{A}$.
Thus, according to Theorem A.4 the equilibrium point x^* is asymptotically
stable.*

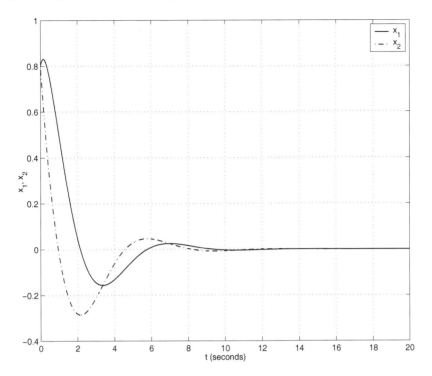

FIGURE A.6
Simulation of (A.7) initialized at $x_1(0) = x_2(0) = 0.8$. Both states converge asymptotically to the equilibrium point $(x_1^*, x_2^*) = (0,0)$.

To visualize the theoretically obtained results, system (A.7) initialized at $(x_1(0), x_2(0)) = (0.8, 0.8)$ *is simulated in Figure A.6. Clearly both states* x_1, x_2 *converge asymptotically to the unique equilibrium point* $(x_1^*, x_2^*) = (0,0)$.

Example A.19 *Consider the system:*

$$\begin{aligned}
\dot{x}_1 &= -x_1 + x_1 x_2 \\
\dot{x}_2 &= -x_1^2.
\end{aligned} \tag{A.8}$$

Notice that the $x^* = (x_1^*, x_2^*) = (0,0)$ *is an equilibrium point of (A.8).*

We shall study its stability properties. In that respect consider the positive definite and radially unbounded function

$$V(x_1, \tilde{x}_2) = \frac{1}{2}x_1^2 + \frac{1}{2}x_2^2.$$

The time derivative of V *along the solutions of (A.8) is*

$$\dot{V} = -x_1^2.$$

Therefore, from Theorem A.2 the solutions x_1, x_2 are uniformly bounded.

To arrive at further conclusions notice that $V(x_1, x_2)$ is bounded from below by $(0,0)$ and is a nonincreasing function of time. Therefore it has a limit as $t \to \infty$ (i.e., $\lim_{t \to \infty} V(x_1(t), x_2(t)) = V(\infty)$). Intergrading $\dot{V}$ we obtain

$$\int_0^t x_1^2(\tau)d\tau = V(0) - V(t)$$

or

$$\lim_{t \to \infty} \int_0^t x_1^2(\tau)d\tau = V(0) - V(\infty) < \infty$$

from which it is concluded that $x_1 \in \mathcal{L}_2$. As x_1, x_2 are uniformly bounded it follows from $\dot{x}_1 = -x_1 + x_1 x_2$ that $\dot{x}_1 \in \mathcal{L}_\infty$. Hence, since $x_1 \in \mathcal{L}_2 \cap \mathcal{L}_\infty$ and $\dot{x}_1 \in \mathcal{L}_\infty$ employing Barbălat's Lemma we conclude that $\lim_{t \to \infty} x_1(t) = 0$.

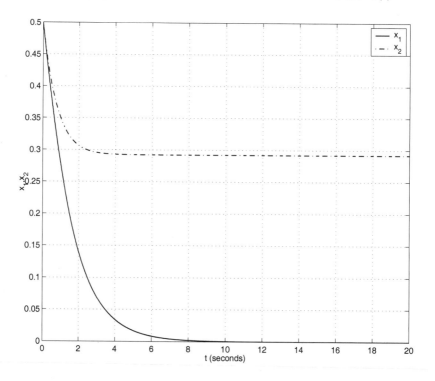

FIGURE A.7

State evolution with respect to time for (A.8). Both states initialize at $(x_1(0), x_2(0)) = (0.5, 0.5)$. The equilibrium $(x_1^*, x_2^*) = (0,0)$ is uniformly stable and additionally x_1 converges asymptotically to zero.

Figure A.7 presents simulations of (A.8). Clearly the equilibrium $(x_1^*, x_2^*) = (0,0)$ is uniformly stable and additionally x_1 converges asymptotically to zero as theory predicts.

Example A.20 *Consider the system:*

$$\begin{aligned}
\dot{x}_1 &= x_1 x_2^2 - x_1(x_1^2 + x_2^2 - 3) \\
\dot{x}_2 &= -x_1^2 x_2 - x_2(x_1^2 + x_2^2 - 3).
\end{aligned} \tag{A.9}$$

Let us choose the Lyapunov function candidate

$$V(x_1, x_2) = \frac{1}{2}x_1^2 + \frac{1}{2}x_2^2$$

which is positive definite and radially unbounded. Its time derivative along the solutions of (A.9) is

$$\dot{V} = -(x_1^2 + x_2^2)(x_1^2 + x_2^2 - 3).$$

Notice that $\dot{V} < 0$ provided that $x_1^2 + x_2^2 > 3$. Therefore, according to Theorem A.2, the solution x_1, x_2 is uniformly ultimately bounded with respect to the set $\{x_1, x_2 \in \Re : x_1^2 + x_2^2 \le 3\}$. Stated otherwise, the solution x_1, x_2 will reach in finite time the neighborhood defined by the circular disk $x_1^2 + x_2^2 \le 3$ with its center the equilibrium point $(x_1^*, x_2^*) = (0, 0)$.

Figure A.8 illustrates the uniform ultimate boundedness property of x_1, x_2. Notice that (A.9) initializes outside the set $\{x_1, x_2 \in \Re : x_1^2 + x_2^2 \le 3\}$ (i.e., $x_1(0) = x_2(0) = 1.5$) and at $t = 1.25$ seconds all system trajectories have entered the circle $x_1^2 + x_2^2 = 3$ from which they never escape.

Example A.21 *Consider the system:*

$$\begin{aligned}
\dot{x}_1 &= x_2 - x_1^3 \\
\dot{x}_2 &= -x_2 - e^{-t}x_1
\end{aligned} \tag{A.10}$$

and choose the positive definite Lyapunov function candidate:

$$V(t, x_1, x_2) = \frac{1}{2}x_1^2 + \frac{1}{2}e^t x_2^2.$$

Its time derivative along the solutions of (A.10) is

$$\begin{aligned}
\dot{V}(t, x) &= \frac{\partial V}{\partial t} + \frac{\partial V}{\partial x_1}\dot{x}_1 + \frac{\partial V}{\partial x_2}\dot{x}_2 \\
&= \frac{1}{2}e^t x_2^2 + x_1 \dot{x}_1 + e^t x_2 \dot{x}_2 \\
&= \frac{1}{2}e^t x_2^2 + x_1[x_2 - x_1^3] + e^t x_2[-x_2 - e^{-t}x_1] \\
&= -\frac{1}{2}e^t x_2^2 - x_1^4.
\end{aligned}$$

Hence, $\dot{V}$ is negative definite. However, according to Theorem A.1, since V is not decrescent, we cannot conclude anything more than the stability of the equilibrium point $x^* = (x_1^*, x_2^*) = (0, 0)$ of (A.10).

Visual verification of the aforementioned theoretical derivations is provided in Figure A.9.

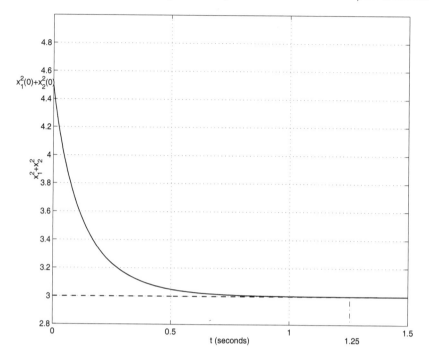

FIGURE A.8

Graphical illustration of the uniform ultimate of (A.9). The states initialize at $x_1(0) = x_2(0) = 5$ and in finite time enter the circle $x_1^2 + x_2^2 = 3$ from which they never escape.

Although Lyapunov's direct method does not reside on the explicit solution of the system's differential equations to characterize the stability of an equilibrium point or the boundedness of its solutions, it suffers from the difficulty of defining a candidate Lyapunov function. Even though in certain cases the existence of such a function is assured, no general procedure exists for its determination.

Stability and convergence results may be derived more conveniently sometimes by employing functions that resemble the properties of the Lyapunov function and for that reason are called Lyapunov-like functions. The following example highlights the approach.

Example A.22 *Consider the system:*

$$
\begin{aligned}
\dot{x}_1 &= x_2^2 x_3^3 \\
\dot{x}_2 &= -x_2 - x_1 x_2 x_3^3 \\
\dot{x}_3 &= x_2^2
\end{aligned}
\tag{A.11}
$$

with initial conditions $x_1(0) = x_{10}, x_2(0) = x_{20}, x_3(0) = x_{30}$. Notice that

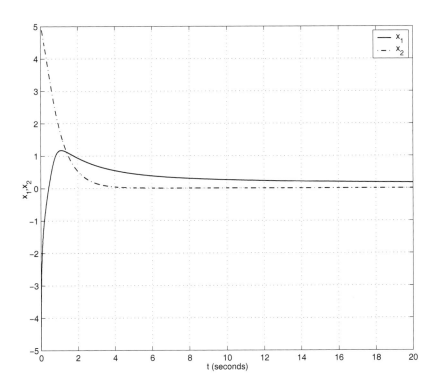

FIGURE A.9

Simulation of system (A.10) with initial conditions $x_1(0) = -5$, $x_2(0) = 5$. From the graphical illustration it is clearly deduced that the equilibrium point $(x_1^*, x_2^*) = (0,0)$ is stable.

(A.11) has nonisolated equilibrium points of the form $(x_1^*, x_2^*, x_3^*) = (c_1, 0, c_2)$ where c_1, c_2 are some constants. Let us now consider the quadratic function

$$V(x_1, x_2) = \frac{1}{2}x_1^2 + \frac{1}{2}x_2^2$$

which is positive semidefinite in $\Re^3$ and therefore does not satisfy the positive definite condition of Theorems A.1-A.4. Thus $V(x_1, x_2)$ is a Lyapunov-like function candidate.

The time derivative of V along the solutions of (A.11) is

$$\dot{V} = -x_2^2 \leq 0 \tag{A.12}$$

which implies that $V(x_1, x_2)$ is a nonincreasing function of time. Hence, it is bounded by:

$$0 \leq V(x_1, x_2) \leq V(x_{10}, x_{20}) = V_0.$$

Therefore, $V, x_1, x_2 \in \mathcal{L}_\infty$. Moreover, V has a limit as $t \to \infty$ i.e.,

$$\lim_{t \to \infty} V(x_1(t), x_2(t)) = V_\infty.$$

Intergrading (A.12) we obtain

$$\int_0^\infty x_2^2(\tau)d\tau = V_0 - V_\infty < \infty.$$

Hence, $x_1 \in \mathcal{L}_2$. Intergrading the last equation of (A.11) we also obtain $x_3 \in \mathcal{L}_\infty$. Therefore, since $x_1, x_2, x_3 \in \mathcal{L}_\infty$ we have $\dot{x}_2 \in \mathcal{L}_\infty$. Since $x_2 \in \mathcal{L}_2 \cap \mathcal{L}_\infty$ and $\dot{x}_2 \in \mathcal{L}_\infty$ employing Barbălat's Lemma we conclude $\lim_{t\to\infty} x_2(t) = 0$.

Summarizing, we have shown using Lyapunov-like functions and extra convergence arguments that the solutions x_1, x_2, x_3 of (A.11) are uniformly bounded and additionally $x_2(t) \to 0$ as $t \to \infty$, for any finite initial condition x_{10}, x_{20}, x_{30}.

System (A.11) was simulated and the results are pictured in Figure A.10. All system states were initialized at 0.8. Clearly, x_2 converges asymptotically to zero, while x_1, x_3 are kept bounded.

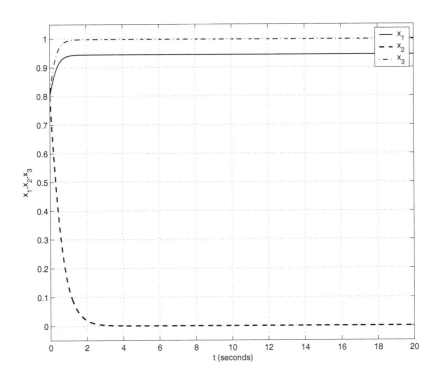

FIGURE A.10

State trajectories of (A.11) with $x_1(0) = x_2(0) = x_3(0) = 0.8$. The x_2 state converges to zero asymptotically, while x_1, x_3 are bounded.

B

Neural Networks for Function Approximation

CONTENTS

Central in the analysis and design of the *NNRC* framework discussed in Chapters 7 - 10 is overcoming the inability of knowing a priori the mathematical expressions relating source transmission rate, with crucial network operation characteristics such as RTT, fairness, link queue size, utilization etc.

Significant aid in this direction was provided by the wide exploitation of parametric approximation structures with neural networks being a specific implementation. In this appendix, details are presented regarding the use of neural networks for function approximation.

B.1   General

We shall begin by defining the problem.

Definition B.1 *(Function Approximation) Given a number of data pairs*

(x_i, y_i), $i = 1, 2, ..., N$ *construct a function that best represents the unknown relationship between* x *and* y *(i.e.,* $y = f(x)$*).*

The solution to a function approximation problem typically follows two steps. In the first, an approximation structure is selected, that is, a certain functional form involving a number of free parameters (the structure selection problem). In the second step, these parameters are appropriately tuned to minimize a distance measure from the unknown function to be approximated (the learning problem). The problems defined in these steps, even though treated independently, are highly correlated, meaning that the best result is in most cases achieved through repetitive applications of both steps until the resulted approximation fulfills the design requirements (e.g., an upper bound on the approximation accuracy). The block diagram of Figure B.1 clarifies the design procedure.

Following the aforementioned discussion, the approximation can be more formally framed as follows. Suppose we are interested in approximating a function $F : \Re^n \to \Re$ in $\Omega \subset \Re^n$, where Ω is some compact subset of $\Re^n$. Let $\xi \in \Re^L$ be a parameter vector and $\hat{F}$ be a known function representing the structure selected to approximate F. Note that $\hat{F}$ is easily evaluated once ξ is fixed. We postulate the functional form:

$$F(x) \simeq \hat{F}(x, \xi). \tag{B.1}$$

After $\hat{F}$ is selected it remains to choose the parameter vector ξ so as to minimize a measure of the distance between the (possibly unknown) function F and its approximation $\hat{F}$. Specific, representative neural network constructions of $\hat{F}$ will be provided in the next section.

The successful pursuit of the function approximation problem is based on the following existence result which is commonly shared by all neural network approximation structures [63], [64], [90], [107], [229], [297].

Density Property: For every continuous function $F : \Re^n \to \Re$ and for every $\varepsilon > 0$, there exist representations $\hat{F}$ and an optimal parameter vector $\xi^* \in \Re^L$ such that

$$\sup_{x \in \Omega} \| F(x) - \hat{F}(x, \xi^*) \| \leq \varepsilon \tag{B.2}$$

where $\Omega \subset \Re^n$ is a compact set.

In (B.2) ε represents the desired approximation accuracy. Following the density property, we may substitute, without loosing generality, any unknown function $F(x)$ by an approximate representation $\hat{F}(x, \xi^*)$, plus a modeling error term $\omega(x)$ obtaining

$$F(x) = \hat{F}(x, \xi^*) + \omega(x), \quad \forall x \in \Omega \subset \Re^n$$

with

$$\| \omega(x) \| \leq \bar{\omega}, \quad \forall x \in \Omega \subset \Re^n$$

where $\bar{\omega}$ is an unknown positive and small constant.

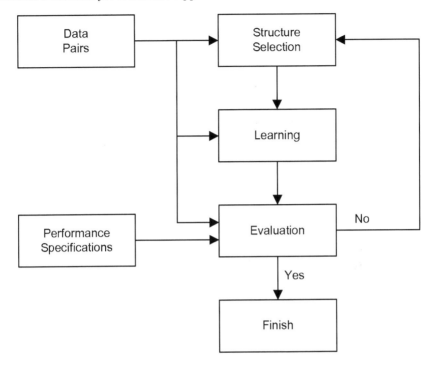

FIGURE B.1
The function approximation design procedure. Given a number of data pairs and strict performance specifications, a candidate structure is first selected and its free parameters are tuned following a learning procedure. The resulted scheme is consequently evaluated to confirm the satisfaction of the application imposed performance specifications. If the output evaluation module is positive the procedure ends. However, in case of a negative evaluation, the whole procedure is repeated by selecting a different approximation structure, performing learning, validating the new result, until receiving a positive evaluation.

This explains the utilization of the density property in the analysis and design of the *NNRC* framework.

B.2   Neural Networks Architectures

Depending on the way the parameters ξ appear in the approximant (neural network) $\hat{F}$, the candidate structures are classified either as linear-in-the-parameters (LIP) or as nonlinear-in-the-parameters (NLIP). Specifically:

$$\hat{F}(x,\xi) \;=\; \xi^T \phi(x) \qquad (LIP - structure) \tag{B.3}$$
$$\hat{F}(x,\xi) \;=\; \theta^T \phi(x,\vartheta) \quad (NLIP - structure) \tag{B.4}$$

where $x \in \Omega \subset \Re^n$, $\hat{F} : \Omega \to \Re$, $\theta \in \Re^{L_1}$, $\vartheta \in \Re^{L_2}$, $\xi \in \Re^L$ with L, L_1, L_2 are positive integers.

Moreover, $\xi = [\theta \; \vartheta]^T$ in (B.4). In addition, the vector ϕ that appears in (B.3), (B.4) represents appropriately selected (nonlinear) regressors that uniquely define each neural network architecture.

As mentioned in the previous section, both LIP and NLIP architectures satisfy the density property. Therefore generality is not harmed. However, the solution to the problem of training a LIP structure to achieve (B.2) is unique and by far less complex, compared to NLIP structures.

The objective of this section is to present several neural network constructions in a unifying manner.

B.2.1　Multilayer Perceptron (MLP)

The multilayered perceptron (MLP) or feedforward neural network is the most widely known and used architecture [34], [118], [217].

Its basic building block is the perceptron. The output of the *i-th* perceptron neuron denoted by z_i is

$$\begin{aligned} z_i \;&=\; s\left(w_0 + w_i^T x\right) \\ &=\; s\left(\vartheta_i^T \bar{x}\right) \end{aligned} \tag{B.5}$$

where $w_i = [w_{i1} \; w_{i2} \; ... \; w_{iL_i}]^T \in \Re^{L_i}$, x is the neural network input $x \in \Re^n$, $\vartheta_i = [w_{i0} \; w_i]^T \in \Re^{L_i+1}$, $\bar{x} = [1 \; x]^T \in \Re^{n+1}$.

The function s in (B.5) is called the activation function and is typically chosen to be of saturation type. Common choices are sigmoid functions such as

$$s(u) = \frac{1}{1 + e^{-u}}$$

and the hyperbolic tangent

$$s(u) = \tanh(u) = \frac{1 - e^{-2u}}{1 + e^{-2u}}.$$

Schematically, the perceptron is pictured in Figure B.2.

Several perceptrons placed in parallel and connected to a single output neuron through the weights $\theta = [\theta_1 \; \theta_2 \; ... \; \theta_{L_i}]^T \in \Re^{L_i}$, form the multilayer perceptron neural network with one hidden layer of neurons.

The activation function of the output layer neurons is typically linear. Therefore, the MLP output y is given by:

$$y = \theta^T s(\vartheta_i^T \bar{x}). \tag{B.6}$$

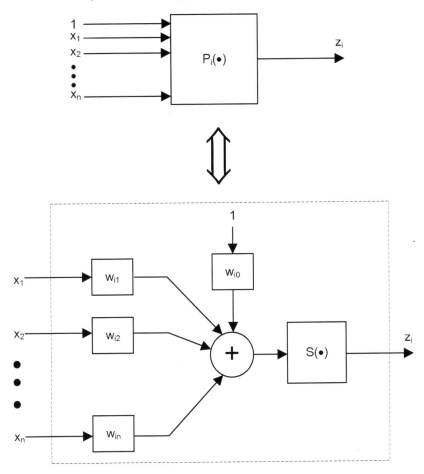

FIGURE B.2

The *i-th* perceptron neuron receiving as inputs x_i, $i = 1, 2, ..., n$.

From a direct comparison of (B.4) and (B.6) we conclude that the MLP is LIP with respect to the output layer weights θ and NLIP with respect to the hidden layer weights ϑ_i. Hence, the MLPs are generally NLIP structures. The MLP neural network is illustrated in Figure B.3.

MLPs with multiple hidden layers have also been reported in the literature. This can be easily accomplished by defining θ as a matrix, making now y to be a vector. Note, however, that the incorporation of additional to one hidden layers significantly increases the complexity.

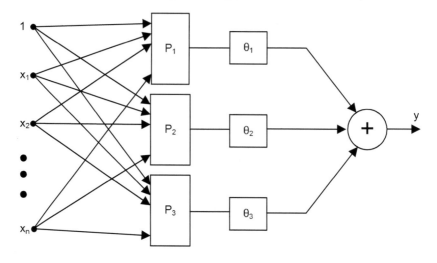

FIGURE B.3
The structure of an MLP having one hidden layer comprised of 3 perceptrons P_i, $i = 1, 2, 3$, n-inputs x_i, $i = 1, 2, ..., n$ and 1 output y.

B.2.2   Radial Basis Function Networks (RBF)

The basic building block of the RBF neural networks is the RBF neuron, whose activation function is comprised by radial basis functions. Originally, radial basis functions were used for interpolation purposes [232]. Their relation to neural networks was provided later (see [43], [86]).

The output of the i-th RBF neuron, denoted by z_i is given by:

$$z_i = g(\|x - c_i\|) \tag{B.7}$$

where $x = [x_1 \; x_2 \; ... \; x_n]^T$ is the input vector to the neuron, $c_i = [c_{i1} \; c_{i2} \; ... \; c_{in}]^T$ is a center vector and $\|x - c_i\|$ is the distance of the input vector to the center locations. In (B.7) $g : \Re_+ \to \Re$ is a radial function with typical representative the gaussian

$$g(u) = e^{-\frac{1}{2}\frac{u_i^2}{\sigma_i^2}} \tag{B.8}$$

with $\sigma_i > 0$ denoting the standard deviation. Schematically, the RBF neuron is pictured in Figure B.4.

Several RBF neurons placed in parallel and connected to a single output neuron of perceptron type with linear activation, through the weights $\theta = [\theta_1 \; \theta_2 \; ... \; \theta_{L_i}]^T \in \Re^{L_i}$, form the RBF neural network with one hidden layer. Therefore, the RBF output y is given by:

$$y = \theta^T g(\|x - c_i\|) \tag{B.9}$$

which is illustrated in Figure B.5.

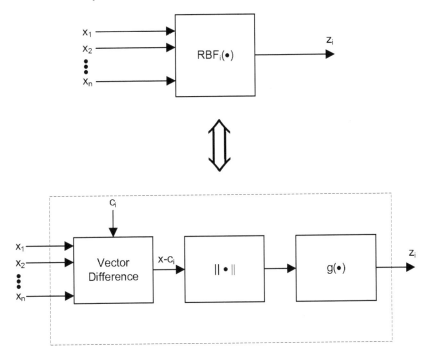

FIGURE B.4
The i-th RBF neuron receiving as inputs x_i, $i = 1, 2, ..., n$.

If the parameters of the gaussian activation function (σ_i, c_i) are predetermined then the only free parameters are the output weights θ. Making the RBF network a LIP structure. If, however, the gaussian function parameters are also free, then the RBF network becomes a NLIP structure.

B.2.3   High-Order Neural Networks (HONN)

High order neural networks (HONNs) are constructed as polynomials of sigmoid functions. Therefore, each basic element (i.e., each high order neuron) implements a single term of the polynomial.

High order networks first appeared within the dynamic neural network configurations of the Hopfield [124] and Cohen-Grossberg [61] models. Their increased storage capacity has been demonstrated in [19, 226], while their stability properties for fixed parameter values have been studied in [67], [155]. Application of HONNs in dynamic systems identification and control can be found in [244].

The output of the i-th high order neuron, denoted by z_i is given by:

$$z_i = \prod_{j \in I_i} [s(x_j)]^{d_j(i)} \tag{B.10}$$

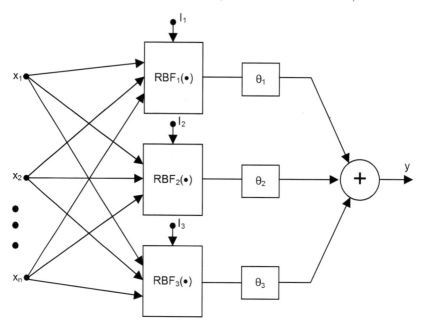

FIGURE B.5
Illustration of a RBF neural network having 3 neurons in the hidden layer, 1 output, receiving n inputs x_i, $i = 1, 2, ..., n$.

where $s(\cdot)$ is a sigmoid function of the form:

$$s(u) = \frac{\alpha}{1 + e^{-\beta(u-c)}} - \gamma \tag{B.11}$$

where the parameters α, β represent the bound and slope of sigmoid's curvature and c, γ are bias constants. In the special case where $\alpha = \beta = 1$, $\gamma = c = 0$ we obtain the logistic function and by setting $\alpha = \beta = 2$, $\gamma = 1$, $c = 0$ we obtain the hyperbolic tangent function.

Furthermore, in (B.10) I_i is a not-ordered subset of $\{1, 2, ..., n\}$ with n being the number of HONN inputs (x_i, $i = 1, 2, ..., n$). Finally, $d_j(i)$ are nonnegative integers. Schematically, the HO neuron is illustrated in Figure B.6.

Several HO neurons placed in parallel and connected to a single output neuron with linear activation function, through the weights $\theta = [\theta_1 \; \theta_2 \; ... \; \theta_{L_i}]^T \in \Re^{L_i}$, form the HONN with one hidden layer. Therefore, the HONN output y is given by:

$$y = \theta^T \prod_{j \in I_i} [s(x_j)]^{d_j(i)} \tag{B.12}$$

and the HONN structure is pictured in Figure B.7.

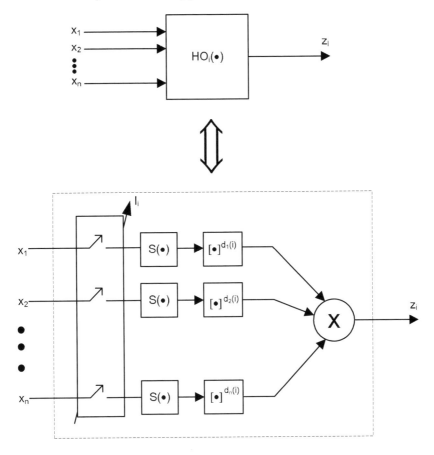

FIGURE B.6

The i-th HO neuron receiving as inputs x_j, $j \in I_i$ with I_i being a not-ordered subset of $\{1, 2, ..., n\}$.

If the parameters of the sigmoid activation functions α, β, γ, c, their powers d as well as the not-ordered subsets I are considered fixed, then the only free parameters are the output weights θ. In that case the HONN is a LIP structure. Otherwise, HONNs become NLIP.

B.3   Off-Line Training

Following the discussion at the beginning of this appendix, off-line training is the problem of minimizing a functional of the distance (objective function) between the possibly unknown input-output map F, relating a given number

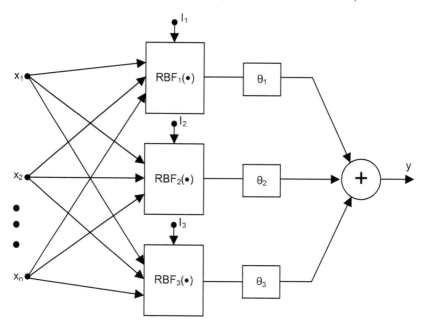

FIGURE B.7

Illustration of a HONN having 3 neurons in the hidden layer, 1 output, receiving a total of n inputs x_i, $i = 1, 2, ..., n$.

of data pairs (x_i, y_i), $i = 1, 2, ..., N$ and its approximant $\hat{F}(x, \xi)$, via selecting the parameter vector ξ.

In this direction, a typical, though not unique, problem statement reads as follows:

$$\min_{\xi \in \Re^L} \frac{1}{2} \sum_{i=1}^{N} \left[y_i - \hat{F}(x_i, \xi) \right]^2. \tag{B.13}$$

Stated otherwise, we want to find the parameter vector $\xi \in \Re^L$ that best matches the given data pairs (x_i, y_i), $i = 1, 2, ..., N$ in the sense of minimizing the sum of squared errors $\left(y_i - \hat{F}(x_i, \xi) \right)^2$.

In (B.13), $\hat{F}$ is assumed to be NLIP. In case of LIP approximants (B.13) becomes:

$$\min_{\theta \in \Re^L} \frac{1}{2} \sum_{i=1}^{N} \left[y_i - \theta^T \phi(x_i) \right]^2. \tag{B.14}$$

For the optimization problems (B.13), (B.14) the following optimality conditions hold [30], [26].

Proposition B.1 *(Necessary Optimality Conditions) Let ξ^* be a local minimum of $J : \Re^L \to \Re_+$ with $J(\xi) = \frac{1}{2} \sum_{i=1}^{m} \left[y_i - \hat{F}(x_i, \xi) \right]^2$, and assume that*

J is continuously differentiable. Then, $\nabla J(\xi^*) = 0$. If in addition J is twice continuously differentiable, then $\nabla^2 J(\xi^*)$ is positive semidefinite.

Proposition B.2 *(Second Order Sufficient Optimality Conditions) Let $J :$ $\Re^L \to \Re_+$ with $J(\xi) = \frac{1}{2} \sum\limits_{i=1}^{m} \left[y_i - \hat{F}(x_i, \xi) \right]^2$, and assume that J is twice continuously differentiable. Suppose that a vector $\xi^* \in \Re^L$ satisfies:*

- $\nabla J(\xi^*) = 0$,

- $\nabla^2 J(\xi^*) > 0$.

Then, ξ^* is a strict local minimum of J, and there exist scalar $\alpha > 0$ and $\varepsilon > 0$ such that

$$J(\xi) \geq J(\xi^*) + \alpha |\xi - \xi^*|^2, \quad \forall \xi \in \Re^L \tag{B.15}$$

with $|\xi - \xi^*| < \varepsilon$.

In case $\hat{F}$ is LIP, the objective function J becomes

$$J(\theta) = \frac{1}{2} \sum_{i=1}^{N} \left[y_i - \theta^T \phi(x_i) \right]^2 \tag{B.16}$$

with $\xi = \theta$. Apparently, (B.16) is convex with respect to θ. The convexity property strengthens significantly the aforementioned propositions as it can be shown that every local minimum is also global, which is also unique. In addition, the necessary condition $\nabla J(\theta^*) = 0$ is also sufficient, if J is convex.

B.3.1   Algorithms

To solve the training problem introduced in (B.13), (or in (B.14)) we seek for an iterative procedure whose solution produces the parameter vector ξ (or θ). If the developed algorithm processes the entire data set before updating ξ, the proposed training methodology is said to belong to the batch type. Otherwise, if the method cycles through blocks of data in sequence and updates the estimate of ξ after each data block is processed, it is called sequential or incremental.

B.3.1.1   Gradient Algorithms

The family of gradient algorithms for unconstrained minimization of a continuously differentiable objective function $J : \Re^L \to \Re_+$, rely on the idea of iterative descent, which works as follows. Starting at some initial point ξ_o, the algorithm successively generates vectors ξ_k, $k = 1, 2, ...$ according to the general formula:

$$\xi_{k+1} = \xi_k + \gamma_k d_k, \quad k = 0, 1, ... \tag{B.17}$$

where $\gamma_k > 0$ is called the stepsize and the vector d_k represents the search direction that satisfies:

$$J(\xi_{k+1}) < J(\xi_k), \quad k = 0, 1, \dots . \tag{B.18}$$

It can be proved [30] that for (B.18) to hold, the direction vector d_k should satisfy

$$\nabla J(\xi_k)^T d_k < 0, \quad k = 0, 1, \dots$$

as long as $\nabla J(\xi_k) \neq 0$, with an "appropriate" selection of the stepsize parameter γ_k. Depending on how d_k is chosen, several well-known methods are produced. Following Propositions B.1, B.2 the methods terminate whenever $\nabla J(\xi_k) = 0$, at which point $\xi_{k+1} = \xi_k$.

Steepest Descent
Selecting

$$d_k = -\nabla J(\xi_k), \quad k = 0, 1, \dots \tag{B.19}$$

and substituting (B.19) into (B.17) the steepest descent method is produced:

$$\xi_{k+1} = \xi_k - \gamma_k \nabla J(\xi_k), \quad k = 0, 1, \dots . \tag{B.20}$$

The method is simple, but its speed of convergence is generally poor.

Newton's Method
If $J(\xi)$ is twice continuously differentiable and its Hessian matrix $\nabla^2 J(\xi_k)$ is positive definite for all $k = 0, 1, \dots$ then selecting

$$d_k = - \left[\nabla^2 J(\xi_k) \right]^{-1} \nabla J(\xi_k), \quad k = 0, 1, \dots \tag{B.21}$$

and substituting (B.21) into (B.17) the Newton's method is obtained:

$$\xi_{k+1} = \xi_k - \gamma_k \left[\nabla^2 J(\xi_k) \right]^{-1} \nabla J(\xi_k), \quad k = 0, 1, \dots . \tag{B.22}$$

The Newton's method with $\gamma_k = 1$ minimizes a positive definite quadratic function in a single iteration. This comes from the fact that it is designed to minimize at each iteration the second order Taylor approximation of J around ξ_k. Generally, the Newton's method is fast converging. However, it requires the calculation of the Hessian $\nabla^2 J(\xi_k)$ and to solve the linear system $\nabla^2 J(\xi_k) d_k = -\nabla J(\xi_k)$ to derive the search direction d_k.

Levenberg–Marquardt Method
The method of Levenberg–Marquardt is a modification to the Newton's method, to ensure descent when the Hessian matrix $\nabla^2 J(\xi_k)$ becomes singular or nearly singular. It is given by the formula:

$$\xi_{k+1} = \xi_k - \gamma_k \left[\nabla^2 J(\xi_k) + \mu_k I \right]^{-1} \nabla J(\xi_k), \quad k = 0, 1, \dots \tag{B.23}$$

where I is the identity matrix and the parameter $\mu_k > 0$ is selected larger than the absolute value of the maximum eigenvalue of $\nabla^2 J(\xi_k)$. In that respect, $\nabla^2 J(\xi_k) + \mu_k I$ is proven to be always positive definite and thus invertible. Hence, (B.23) is valid $\forall k$.

Notice that when μ_k is sufficiently large, the term $\mu_k I$ in $\nabla^2 J(\xi_k) + \mu_k I$ dominates. As a consequence, the Levenberg–Marquardt method behaves almost like a steepest descent. On the contrary, when μ_k is small, then the algorithm practically operates as a Newton method, achieving fast convergence.

Quasi-Newton Methods
Selecting

$$d_k = -\Delta_k \nabla J(\xi_k), \quad k = 0, 1, \ldots \tag{B.24}$$

where Δ_k is a positive definite matrix, and substituting (B.24) into (B.17), we obtain the family of Quasi-Newton methods depending on the explicit selection of Δ_k:

$$\xi_{k+1} = \xi_k - \gamma_k \Delta_k \nabla J(\xi_k). \tag{B.25}$$

Successful methods in this family choose Δ_k to be an approximation of $\left[\nabla^2 J(\xi_k) \right]^{-1}$. Diagonal approximations that simply disregard the off-diagonal element of $\nabla^2 J(\xi_k)$ are very popular alternatives. A more sophisticated implementation of Δ_k, relies on the iterative formula:

$$\Delta_{k+1} = \Delta_k + \frac{p_k p_k^T}{p_k^T q_k} - \frac{\Delta_k q_k q_k^T \Delta_k}{q_k^T \Delta_k q_k} + \phi_k \tau_k u_k u_k^T \tag{B.26}$$

where

$$u_k = \frac{p_k}{p_k^T q_k} - \frac{\Delta_k q_k}{\tau_k} \tag{B.27}$$

$$\tau_k = q_k^T \Delta_k q_k \tag{B.28}$$

$$p_k = \xi_{k+1} - \xi_k \tag{B.29}$$

$$q_k = \nabla J(\xi_{k+1}) - \nabla J(\xi_k) \tag{B.30}$$

with $0 \le \phi_k \le 1$, $\forall k$, and Δ_0 is an arbitrary positive definite matrix. It can be proven [30] that the above-mentioned iterative procedure produces positive definite matrices Δ_k provided γ_k is selected to satisfy

$$\nabla J(\xi_k)^T d_k < \nabla J(\xi_{k+1})^T d_k.$$

Furthermore, the Quasi-Newton methods (B.26)-(B.30) minimize an n-dimensional positive definite quadratic objective function of the form

$$J(\xi) = \frac{1}{2} \xi^T Q \xi - b^T \xi, \quad Q > 0$$

in at most n steps and

$$\Delta_n = \left[\nabla^2 J(\xi_n) \right]^{-1} = Q^{-1}.$$

With respect to convergence rate, Quasi-Newton methods are faster than steepest descent, which converges asymptotically and slower than Newton, which in case of positive definite quadratic objective functions requires a single iteration.

Stepsize Selection

There is a large variety of stepsize selection rules reported in the literature. In this subsection we present some of them.

Minimization

According to the minimization rule, γ_k in each iteration minimizes the objective function J along the direction d_k. In other terms

$$J(\xi_{k+1}) = J(\xi_k + \gamma_k d_k) = \min_{\gamma \geq 0} J(\xi_k + \gamma d_k).$$

A slight modification that is used in many practical problems is to constrain the search interval to satisfy $0 \leq \gamma \leq \gamma_{max}$ with γ_{max} a positive predefined constant. Now γ_k is selected as the value of γ to yield the minimum $J(\xi_k + \gamma d_k)$ in $[0, \gamma_{max}]$. Hence,

$$J(\xi_{k+1}) = J(\xi_k + \gamma_k d_k) = \min_{0 \leq \gamma \leq \gamma_{max}} J(\xi_k + \gamma d_k).$$

The aforementioned modification is referred to as the limited minimization rule. All minimization rules are implemented by incorporating a line search method. Further details may be found in [30], [26]. Owing to their fairly complicated nature, minimization rules for stepsize selection are typically avoided in computationally demanding problems such as neural network training.

Constant stepsize

It is the simplest stepsize rule according to which γ_k is kept constant $\forall k$ to a fixed predetermined value $\gamma > 0$. Hence,

$$\gamma_k = \gamma, \quad k = 0, 1, \dots .$$

Special care should be devoted when selecting γ. If γ is chosen too large, the training algorithm will most probably diverge. On the contrary, if γ is selected too small, the convergence rate of the training algorithm may be very low.

Diminishing stepsize

According to this rule γ_k should converge to zero as k increases, following a predefined reduction pattern. The method is simple. However, it may lead to premature training if γ_k diminishes fast enough, forcing the training algorithm to finish even far from an optimal solution. For this reason it is typically required

$$\sum_{k=0}^{\infty} \gamma_k = \infty.$$

Theoretically the method converges. In practice, though, the convergence rate tends to be slow. Therefore, diminishing stepsize is mainly used for applications where the convergence speed is not the dominant issue, such as in singular problems.

B.3.1.2 Least Squares

Least squares is a training methodology that applies to LIP approximants, solving optimization problems formulated as (B.14).

Given a number of N input-output data pairs (x_i, y_i), $i = 1, 2, ..., N$ and a regressor vector Φ, we define:

$$y = \begin{bmatrix} y_1 & y_2 & \cdots & y_N \end{bmatrix}^T \in \Re^N$$

$$\Phi = \begin{bmatrix} \phi_1(x_1) & \phi_2(x_1) & \cdots & \phi_L(x_1) \\ \phi_1(x_2) & \phi_2(x_2) & \cdots & \phi_L(x_2) \\ \vdots & \vdots & \ddots & \vdots \\ \phi_1(x_N) & \phi_2(x_N) & \cdots & \phi_L(x_N) \end{bmatrix} \in \Re^{N \times L}.$$

Hence, in matrix form (B.14) becomes

$$\min_{\theta \in \Re^L} \frac{e^T e}{2} \tag{B.31}$$

where

$$e = y - \Phi\theta.$$

Owing to the convexity with respect to θ of the objective function $\frac{e^T e}{2}$, there exists a single solution to the problem, which satisfies:

$$\frac{d}{d\theta} \frac{|y - \Phi\theta|^2}{2} = 0$$

from which we arrive at the normal equations

$$\theta^T \Phi^T \Phi = y^T \Phi. \tag{B.32}$$

If $\Phi^T \Phi$ is nonsingular, its inverse exists yielding

$$\theta^T = y^T \Phi (\Phi^T \Phi)^{-1}. \tag{B.33}$$

To avoid matrix inversion, the parameters θ are preferably determined via solving the normal equation (B.32) rather than (B.33).

To obtain a geometric interpretation of the operation of the least squares algorithm, let Φ be expressed as a row of L columns of size $N \times 1$ as follows:

$$\Phi = \begin{bmatrix} \phi_1 & \phi_2 & \cdots & \phi_L \end{bmatrix}.$$

Then

$$\Phi\theta = \begin{bmatrix} \phi_1 & \phi_2 & \cdots & \phi_L \end{bmatrix} \begin{bmatrix} \theta_1 \\ \theta_2 \\ \vdots \\ \theta_L \end{bmatrix}.$$

In other terms, $\Phi\theta$ is a linear combination of the basis $\{\phi_1, \phi_2, ..., \phi_L\}$ in an L-dimensional space. Thus, for $e = y - \Phi\theta$ to achieve a minimum length, $\Phi\theta$ must be equal to the projection of y onto the plane spanned by $\{\phi_1, \phi_2, ..., \phi_L\}$.

B.3.1.3   Backpropagation

The backpropagation algorithm is the application of steepest descent on the MLP neural networks. It became very popular after the publication of [248], even though it was proposed many years earlier by Werbos [295].

The input-output relation of a MLP neural network having n inputs, one output and one hidden layer of L neurons, is given by

$$z_i = s\left(w_0 + \sum_{j=1}^{n} w_{ij}x_j\right), \quad i = 1, 2, ..., L$$

$$\hat{y} = \sum_{i=1}^{L} \theta_i z_i$$

where x_j is the $j - th$ input, $\hat{y}$ is the output θ_i, w_0, w_{ij}, $i = 1, 2, ..., L$, $j = 1, 2, ..., n$ are neural network weights and s is the neuron activation function typically expressed by

$$s(u) = \frac{1}{1 + e^{-u}}.$$

Given a desired output value y, the problem is to minimize the quadratic error objective function

$$J = \frac{1}{2}e^2$$
$$= \frac{1}{2}(\hat{y} - y)^2.$$

Applying the steepest descent rule to the i-th weight of the output layer θ_i we have:

$$\theta_{ik+1} = \theta_{ik} - \gamma \frac{\partial J}{\partial \theta_i}$$
$$= \theta_{ik} - \gamma e_k \frac{\partial \hat{y}}{\partial \theta_i}$$
$$= \theta_{ik} - \gamma e_k z_i$$

where k denotes the iteration number.

The update laws of the weights of the hidden layer w_{ij} are similarly obtained. However, now

$$w_{ij}^{k+1} = w_{ij}^k - \gamma \frac{\partial \hat{y}}{\partial w_{ij}}$$

where $\frac{\partial \hat{y}}{\partial w_{ij}}$ is obtained by applying the chain rule

$$\frac{\partial \hat{y}}{\partial w_{ij}} = \frac{\partial \hat{y}}{\partial z_i} \frac{\partial z_i}{\partial u_i} \frac{\partial u_i}{\partial w_{ij}}$$

where

$$u_i = w_0 + \sum_{j=1}^{n} w_{ij} x_j, \quad i = 1, 2, ..., n$$

and

$$\frac{\partial \hat{y}}{\partial z_i} = \theta_i$$

$$\frac{\partial z_i}{\partial u_i} = s(u_i)(1 - s(u_1))$$

$$\frac{\partial u_i}{\partial w_{ij}} = x_j.$$

B.4 On-Line Training

In off-line training discussed thus far, an input-output data set is first acquired and then the free parameters of a neural network approximant are selected to best fit the data. On the other hand, in on-line training, an initial parameter estimate is first chosen and then estimates are recursively produced based on current measurements. Hence, the key difference between off-line and on-line training is that in the latter, parameter updates become available in real-time and no data storage is required. The on-line training philosophy is preferable for very large data sets to reduce computational cost and avoid excessive delays.

B.4.1 Filtering Schemes

Consider the problem of approximating the continuous though unknown nonlinear mapping $f : \Re^n \to \Re$, representing the input-output relation $y = f(x)$, in $\Omega \subset \Re^n$ where Ω is some compact set. To continue, we construct the filter:

$$\dot{z} = y$$
$$= f(x). \tag{B.34}$$

According to the neural network density property, the unknown f can be expressed as

$$f(x) = \hat{f}(x, \xi) + w(x), \quad \forall x \in \Omega \subset \Re^n \tag{B.35}$$

with

$$\|w(x)\| \leq \bar{\omega}, \quad \forall x \in \Omega \subset \Re^n \tag{B.36}$$

where $\bar{\omega}$ is a positive, unknown but small constant. Substituting (B.35) into (B.34) we obtain

$$\dot{z} = \hat{f}(x, \xi) + w(x), \quad \forall x \in \Omega \subset \Re^n. \tag{B.37}$$

Since in most applications $\dot{z}$ is unavailable for measurement, we filter both sides of (B.37) by a stable first order filter $\frac{1}{s+\lambda}$, $\lambda > 0$ to obtain:

$$J(t) = \frac{1}{s + \lambda} \hat{f}(x, \xi) + \frac{1}{s + \lambda} w(x) \tag{B.38}$$

where

$$J(t) = \frac{1}{s + \lambda} z. \tag{B.39}$$

Note that $J(t)$ is obtained by solving:

$$\left. \begin{array}{rcl} \dot{J} & = & -\lambda J + z \\[2mm] \dot{z} & = & y \end{array} \right\}. \tag{B.40}$$

Defining

$$\delta(t) = \frac{1}{s + \lambda} w(x)$$

(B.38) becomes

$$J(t) = \frac{1}{s + \lambda} \hat{f}(x, \xi) + \delta(t). \tag{B.41}$$

Hence, since $w(x)$ is small in the sense of (B.36), $\delta(t)$ is also small and $\hat{f}$ dominates (B.41). What we have achieved in this way is to transform the type of uncertainty from structural (unknown $f(x)$) to parametric (unknown ξ).

B.4.1.1   Filtered Error

Following (B.38), to construct the filtered error scheme we first propose the neural network estimator

$$\hat{J}(t) = \frac{1}{s + \lambda} \hat{f}(x, \hat{\xi}) \tag{B.42}$$

obtained by replacing the unknown parameters (neural network weights) ξ by their estimates $\hat{\xi}$. We also define the output estimation error

$$e(t) = \hat{J}(t) - J(t) \tag{B.43}$$

where $J(t)$ is obtained by (B.40). Notice also that $\hat{J}(t)$ is provided by

$$\dot{\hat{J}} = -\lambda \hat{J} + \hat{f}(x, \hat{\xi}).$$ (B.44)

Substituting (B.41), (B.42) we get

$$
\begin{aligned}
e(t) &= \frac{1}{s+\lambda} \left[\hat{f}(x, \hat{\xi}) - \hat{f}(x, \xi) \right] - \delta(t) \\
&= \frac{1}{s+\lambda} \left[\hat{f}(x, \hat{\xi}) - f(x) \right].
\end{aligned}
$$ (B.45)

From (B.45) we conclude that $e(t)$ is equal to filtering the approximation error $\hat{f}(x, \hat{\xi}) - f(x)$; thus the term filtered error. Depending on how large λ is, the output estimation error $e(t)$ follows closely the approximation error $\hat{f}(x, \hat{\xi}) - f(x)$. However, extremely large values of λ make the scheme highly sensitive to the presence of additive measurement noise, which is directly multiplied by λ.

From (B.45) we can easily derive the state-space expression

$$\dot{e} = -\lambda e + \hat{f}(x, \hat{\xi}) - y, \quad \lambda > 0.$$ (B.46)

Equation (B.46) is general and includes both LIP and NLIP structures (B.3) and (B.4), respectively. More specifically, in case of LIP neural networks (B.46) becomes

$$\dot{e} = -\lambda e + \hat{\theta}^T \phi(x) - y, \quad \lambda > 0, \quad \hat{\theta}, \phi \in \Re^L$$ (B.47)

while the NLIP case becomes

$$\dot{e} = -\lambda e + \hat{\theta}^T \phi(x, \hat{\vartheta}) - y, \quad \lambda > 0, \quad \hat{\theta}, \phi \in \Re^{L_0}, \quad \hat{\vartheta} \in \Re^{L_1}.$$ (B.48)

B.4.1.2 Filtered Regressor

The filtered regressor scheme is defined for LIP neural network structures and is constructed based on the estimator

$$\hat{J}(t) = \hat{\theta}^T(t) \eta(t)$$ (B.49)

where $\eta(t)$ is obtained from $\phi(x)$ after filtering:

$$\eta(t) = \frac{1}{s+\lambda} \phi(x).$$ (B.50)

The output estimation error is defined as

$$
\begin{aligned}
e(t) &= \hat{J}(t) - J(t) \\
&= \hat{\theta}^T(t) \eta(t) - \theta^T(t) \eta(t) - \delta(t) \\
&= \tilde{\theta}^T(t) \eta(t) - \delta(t)
\end{aligned}
$$ (B.51)

where $\tilde{\theta} = \hat{\theta} - \theta$. Clearly (B.51) is a static relationship, comparing to the dynamic (B.45) of the filtered error scheme. In state-space (B.49), (B.50) take the form:

$$\dot{\eta} = -\lambda\eta + \phi(x) \tag{B.52}$$

$$\hat{J}(t) = \hat{\theta}^T(t)\eta(t). \tag{B.53}$$

B.4.2   Lyapunov-Based Training

Lyapunov-based training is built upon the filtered error scheme following the Lyapunov synthesis method, which is briefly reported in Appendix A. The most significant advantage is its inherent stability property. In this section we shall consider both LIP and NLIP cases, and we shall assume that no modeling error exists (i.e., $\delta(t) = 0$).

B.4.2.1   LIP Case

Since we assume $\delta(t) = 0$, (B.47) becomes

$$\dot{e} = -\lambda e + \tilde{\theta}^T \phi(x), \quad \lambda > 0, \quad \tilde{\theta} \in \Re^L. \tag{B.54}$$

To continue, consider the Lyapunov function candidate

$$V(e, \tilde{\theta}) = \frac{1}{2}e^2 + \tilde{\theta}^T \Gamma^{-1} \tilde{\theta} \tag{B.55}$$

where Γ is a positive definite matrix of constant gains typically selected as

$$\Gamma = diag\{ \begin{array}{cccc} \gamma_1 & \gamma_2 & \cdots & \gamma_L \end{array} \}.$$

Taking the time derivative of V along the solutions of (B.54) we obtain

$$\dot{V} = -\lambda e^2 + e\tilde{\theta}^T \phi(x) + \tilde{\theta}^T \Gamma^{-1} \dot{\hat{\theta}}.$$

Selecting the training scheme

$$\dot{\hat{\theta}} = \Gamma\phi(x)e \tag{B.56}$$

$\dot{V}$ becomes

$$\dot{V} = -\lambda e^2 \leq 0$$

from which following a standard procedure (see Appendix A for details), certain stability results are deduced.

B.4.2.2   NLIP Case

Assuming $\delta(t) = 0$, (B.48) becomes

$$\dot{e} = -\lambda e + \hat{\theta}^T \phi(x, \hat{\vartheta}) - \theta^T \phi(x, \vartheta), \quad \theta \in \Re^{L_0}, \quad \vartheta \in \Re^{L_1}. \tag{B.57}$$

Using the Taylor series expansion

$$\theta^T \phi(x, \hat{\vartheta}) - \theta^T \phi(x, \vartheta) = \tilde{\theta}^T \phi(x, \hat{\vartheta}) + \tilde{\vartheta} \frac{\partial \phi}{\partial \vartheta}(x, \hat{\vartheta})\theta + \mathcal{K}(x, \tilde{\theta}, \tilde{\vartheta}) \qquad \text{(B.58)}$$

where $\tilde{\theta} = \hat{\theta} - \theta$, $\tilde{\vartheta} = \hat{\vartheta} - \vartheta$ and $\mathcal{K}$ is a term containing the high-order components of the Taylor series expansion. After dropping all high-order terms and substituting (B.58) into (B.57) we obtain

$$\dot{e} = -\lambda e + \tilde{\theta}^T \phi(x, \hat{\vartheta}) + \tilde{\vartheta}^T \frac{\partial \phi}{\partial \vartheta}(x, \hat{\vartheta})\theta. \qquad \text{(B.59)}$$

Let us now consider the Lyapunov function candidate

$$V(e, \tilde{\theta}, \tilde{\vartheta}) = \frac{1}{2}e^2 + \frac{1}{2}\tilde{\theta}^T \Gamma_1^{-1} \tilde{\theta} + \frac{1}{2}\tilde{\vartheta}^T \Gamma_2^{-1} \tilde{\vartheta} \qquad \text{(B.60)}$$

where Γ_i, $i = 1, 2$ are positive definite constant matrices, typically selected as

$$\Gamma_1 = diag\{ \gamma_{11} \quad \gamma_{12} \quad \cdots \quad \gamma_{1L_0} \},$$

$$\Gamma_2 = diag\{ \gamma_{21} \quad \gamma_{22} \quad \cdots \quad \gamma_{2L_1} \}.$$

Differentiating V with respect to time along the solutions of (B.59) we obtain

$$\dot{V} = -\lambda e^2 + e\tilde{\theta}^T \phi(x, \hat{\vartheta}) + e\tilde{\vartheta}^T \frac{\partial \phi}{\partial \vartheta}(x, \hat{\vartheta})\theta + \tilde{\theta}^T \Gamma_1^{-1} \dot{\tilde{\theta}} + \tilde{\vartheta}^T \Gamma_2^{-1} \dot{\tilde{\vartheta}}. \qquad \text{(B.61)}$$

After selecting the training scheme

$$\dot{\hat{\theta}} = \Gamma_1 \phi(x, \hat{\vartheta})e \qquad \text{(B.62)}$$

$$\dot{\hat{\vartheta}} = \Gamma_2 \frac{\partial \phi}{\partial \vartheta}(x, \hat{\vartheta})\theta e \qquad \text{(B.63)}$$

$\dot{V}$ becomes

$$\dot{V} = -\lambda e^2 \le 0.$$

Again, following standard arguments (see Appendix A for details) certain stability results can be obtained.

B.4.3 Steepest Descent Training

The steepest descent training method uses the filtered regressor scheme (B.51) trying to minimize the quadratic cost functional

$$J(\hat{\theta}) = \frac{1}{2}e^2. \qquad \text{(B.64)}$$

The method will be presented in case of no modeling error (i.e., $\delta(t) = 0$).

Employing (B.51), $J(\hat{\theta})$ becomes

$$J(\hat{\theta}) = \frac{1}{2}(\hat{\theta}^T \eta(t) - J(t))^2 \tag{B.65}$$

where $J(t)$ is derived from (B.40). According to the steepest descent algorithm, the parameter estimates $\hat{\theta}$ are updated as follows:

$$\dot{\hat{\theta}} = -\gamma \nabla J(\hat{\theta}) \tag{B.66}$$

where $\gamma > 0$ is a design constant. After obtaining the ∇J from (B.65) we finally arrive at

$$\dot{\hat{\theta}} = -\gamma \eta(t) e(t). \tag{B.67}$$

In theory, the adaptive gain γ in (B.67) may admit any positive value. In practice, however, low values of γ may slow down adaptation, while high values may result in oscillations or even unstable behavior especially in the presence of noise or modeling errors.

B.4.4 Recursive Least Squares Training

The recursive least squares method also uses the filtered regressor scheme (B.51). However, now we try to minimize the cost function:

$$J(\hat{\theta}) = \int_0^t (\hat{\theta}\eta(\tau) - J(\tau))^2 d\tau.$$

As in the steepest descent training, in this method also we consider the no modeling error case $\delta(t) = 0$. The recursive least squares algorithm for the parameter vector $\hat{\theta}$ can be shown to be [137]:

$$\dot{\hat{\theta}} = -P(t)\eta(t)e(t), \quad \hat{\theta}(0) = \hat{\theta}_0 \tag{B.68}$$
$$\dot{P} = -P(t)\eta(t)\eta^T(t)P(t), \quad P(0) = P_0 \tag{B.69}$$

where P is a square matrix called the covariance matrix that initially is chosen to be positive definite. In the above formulation, the least squares method can be thought of as a steepest descent algorithm with a time-varying gain γ.

It can be proven that in the (B.68), (B.69) formulation, $P(t)$ becomes arbitrarily small, thus slowing down adaptation in some directions of the $\hat{\theta}$ vector. This is the covariance wind-up problem. To prevent its appearance, several modifications are reported in the literature [137] including the covariance resetting, the least squares with forgetting factor and the projection method.

A General Comment For all on-line training schemes specific stability properties can be established mainly by exploiting Lyapunov stability theory. However, even if the output estimation error $e(t)$ can be seen to converge asymptotically to zero (when $\delta(t) = 0$), the parameter estimation vector $\hat{\theta}(t)$ may not even approach the optimal parameter vector θ. For this to happen an extra condition is required to hold.

Take, for example, the steepest descent training algorithm:

$$\dot{\hat{\theta}} = -\gamma \eta(t) e(t). \tag{B.70}$$

Substituting $e(t) = \eta^T(t)\tilde{\theta}(t)$ in B.70 we obtain

$$\dot{\hat{\theta}} = -\gamma \eta(t) \eta^T(t) \tilde{\theta}(t).$$

Therefore, the convergence of $\tilde{\theta}(t)$ to zero, requires $\eta(t)\eta^T(t)$ to stay away from zero in some sense. This is the so-called persistency of excitation (PE) condition.

Definition B.2 *(PE Condition) A bounded vector signal $\eta(t) \in \Re^L$ is persistently exciting if there exists $a > 0$ and $T > 0$ such that*

$$\int_t^{t+T} \eta(\tau)\eta^T(\tau)d\tau \geq aI, \quad \forall t \geq 0.$$

B.4.5 Robust On-Line Training

Thus far we have presented training algorithms that were derived under the assumption of no modeling error (i.e., $\delta(t) = 0$), or, otherwise stated, when the unknown nonlinear function $f(x)$ can be exactly represented by the approximant $\hat{f}(x, \xi)$ for some unknown parameter vector $\xi \in \Re^L$. In practice, however, such a property is less likely to hold. It has been shown [137] that even for linear systems, the presence of small modeling errors may drive the parameter estimates $\hat{\xi}$ to infinity (parameter drift phenomenon). To avoid the appearance of this problem various modifications have been reported in the adaptive parameter estimation literature [137]. In this section we shall briefly present the most representative ones. To help presentation, we shall consider throughout the steepest descent algorithm:

$$\dot{\hat{\theta}} = -\Gamma \eta(t) e(t). \tag{B.71}$$

σ-Modification
According to σ-modification (B.71) becomes:

$$\dot{\hat{\theta}} = -\sigma\hat{\theta} - \Gamma \eta(t) e(t). \tag{B.72}$$

Qualitatively, if $\hat{\theta}$ starts drifting to infinity, the term $-\sigma\hat{\theta}$ shall dominate forcing $\hat{\theta}$ to reverse motion. From another perspective, (B.71) is a pure integrator.

Therefore even a small constant error e is capable of driving $\hat{\theta}$ to infinity. In σ-modification, the term $-\sigma\hat{\theta}$ results in a practical integrator implementation, according to which $\hat{\theta}$ saturates and is thus prevented from drifting to infinity.

The advantage of σ-modification is that it does not require knowledge of an upper bound on the modeling error $\delta(t)$. However, it destroys some of the convergence properties obtained in the ideal case ($\delta(t) = 0$).

ε-Modification

The ε-modification was developed as an attempt to eliminate some of the drawbacks of the σ-modification. It is given by:

$$\dot{\hat{\theta}} = -\Gamma|e(t)|\hat{\theta} - \Gamma\eta(t)e(t). \tag{B.73}$$

The idea behind (B.73) is to remove the σ-action (the term $-\sigma\hat{\theta}$ in (B.72)) whenever $e(t) = 0$, where no modification is actually needed in (B.72)). When the ε-modification term is activated, the method behaves like σ-modification.

Dead-Zone Modification

When $\delta(t) = \frac{1}{s+\lambda}\omega \neq 0$ it can be seen, following Lyapunov analysis, that the time derivative of the Lyapunov function is negative semidefinite for $|e| > |\omega|$. When $|e| < |\omega|$ the parameter estimates may diverge and the Lyapunov function may increase. A simple solution to increase robustness is to stop $\hat{\theta}$ adaptation when $|e| < |\omega|$. This results in dead-zone modification, which is given by:

$$\dot{\hat{\theta}} = \begin{cases} -\Gamma\eta(t)e(t) & , \text{ if } |e(t)| \geq \bar{\omega} \\ 0 & , \text{ otherwise} \end{cases} \tag{B.74}$$

where $\bar{\omega} > 0$ is an upper bound on ω. Knowledge of $\bar{\omega} > 0$ is a drawback of the method. Another disadvantage of the dead-zone is that even in the case where $\omega = 0$ asymptotic stability cannot be proved. However, uniform ultimate boundedness of e is obtained with respect to a set whose size is determined by $\bar{\omega}$ and some design parameters.

Projection Modification

Provided we have knowledge of a convex set $\mathcal{S}$ where the optimal parameter vector θ belongs, the projection modification method can be used to prevent parameter drift, provided we initialize from inside $\mathcal{S}$ (i.e., $\hat{\theta}(0) \in \mathcal{S}$).

The idea of projection modification reads as follows: start with $\hat{\theta}(0) \in \mathcal{S}$ and update $\hat{\theta}$ according to (B.71) until $\hat{\theta}$ is on the boundary $\partial\mathcal{S}$ of $\mathcal{S}$ and is intended to move outwards. Then the derivative of $\hat{\theta}(t)$ is projected onto the tangent to $\partial\mathcal{S}$ hyperplane.

Let the convex set $\mathcal{S}$ be defined via

$$\mathcal{S} = \{\theta \in \Re^L : g(\theta) \leq 0\}$$

where $g : \Re^L \to \Re$ is a smooth function. Let us also denote with $\mathcal{S}_{in}$ the interior of $\mathcal{S}$. The projection modification is provided by [137]:

$$\dot{\hat{\theta}} = \begin{cases} -\Gamma\eta(t)e(t) & , \text{if } \hat{\theta} \in \mathcal{S}_{in} \text{ or if } \hat{\theta} \in \partial\mathcal{S} \\ & \text{and } \nabla g^T \Gamma\eta(t)e(t) \geq 0 \\ \\ -\Gamma\eta(t)e(t) + \Gamma\frac{\nabla g \nabla g^T}{g^T\Gamma\nabla g}\Gamma\eta(t)e(t) & , \text{otherwise} \end{cases} \quad . \quad (B.75)$$

The projection modification algorithm keeps the parameter estimates always within the convex set $\mathcal{S}$, without destroying the stability properties obtained in its absence.

Bibliography

[1] Ramachandra Achar and Michel S. Nakhla. Simulation of high speed interconnects. *Proceedings of the IEEE*, 89(5):693–728, 2001.

[2] Micah Adler, Jin Yi Cai, Jonathan Shapiro, and Don Towsley. Estimation of congestion price using probabilistic packet marking. In *Proceedings of the IEEE International Conference on Computer Communications 22nd Annual Joint Conference of the IEEE Computer and Communications Societies*, pages 2068–2078, April 2003.

[3] Ishtiaq Ahmed, Okabe Yasuo, and Kanazawa Masanori. Improving performance of SCTP over broadband high latency networks. In *Proceedings of 28th Annual IEEE International Conference on Local Computer Networks*, pages 644–645, October 2003.

[4] Jay Aikat, Jasleen Kaur, F. Donelson Smith, and Kevin Jeffay. Variability in TCP round-trip times. In *Proceedings of the 3rd ACM Special Interest Group on Data Communication Conference on Internet Measurement*, pages 279–284, October 2003.

[5] Ozgur B. Akan, Jian Fang, and Ian F. Akyildiz. Performance of TCP protocols in deep space communication networks. *IEEE Communications Letters*, 6(11):478–480, 2002.

[6] Mark Allman, Sally Floyd, and Craig Partridge. Increasing TCP's initial window. *RFC 3390, Proposed Standard*, 2002.

[7] Mark Allman, Vern Paxson, and Richard Stevens. TCP Congestion Control. *RFC 2581*, April 1999.

[8] Tansu Alpcan and Tamer Başar. A globally stable adaptive congestion control scheme for Internet-style networks with delay. *IEEE/ACM Transactions on Networking*, 13(6):1261–1274, 2005.

[9] Marios C. Angelides and Harry Agius. *The Handbook of MPEG Applications: Standards in Practice*. John Wiley and Sons, New York, 2011.

[10] Søren Asmussen. *Applied Probability and Queues (Stochastic Modelling and Applied Probability)*. Springer-Verlag, New York, 2003.

[11] Karl Johan Astrom and Bjorn Wittenmark. *Adaptive Control (2nd Edition)*. Addison-Wesley Longman Publishing Co., Boston, MA, 1994.

[12] Sanjeewa Athuraliya, Steven H. Low, Victor H. Li, and Qinghe Yin. REM: active queue management. *IEEE Network*, 15(3):48–53, May 2001.

[13] David Austerberry. *The Technology of Video and Audio Streaming, Second Edition*. Focal Press, Burlington, MA, 2004.

[14] James Aweya, Michel Ouellette, and Delfin Y. Montuno. Relative loss rate differentiation: performance of short-lived TCP flows: Research articles. *International Journal of Communication Systems*, 18(1):77–93, February 2005.

[15] François Baccelli, Giovanna Carofiglio, and Marta Piancino. Stochastic analysis of scalable TCP. In *Proceedings of the 31st IEEE International Conference on Computer Communications*, pages 19–27, April 2009.

[16] Hamsa Balakrishnan, Nandita Dukkipati, Nick McKeown, and Claire J. Tomlin. Stability analysis of explicit congestion control protocols. *IEEE Communications Letters*, 11(10):823–825, 2007.

[17] Hari Balakrishnan, Venkata N. Padmanabhan, and Randy H. Katz. The effects of asymmetry on TCP performance. *Mobile Networks and Applications*, 4(3):219–241, 1999.

[18] Mark Balch. *Complete Digital Design: A Comprehensive Guide to Digital Electronics and Computer System Architecture (Professional Engineering)*. McGraw-Hill, New York, 2003.

[19] Pierre Baldi. Neural networks, orientations of the hypercube and algebraic threshold functions. *IEEE Transactions on Information Theory*, 34(3):523–530, 1988.

[20] Tarun Banka, Abhijit A. Bare, and Anura P. Jayasumana. Metrics for degree of reordering in packet sequences. In *Proceedings of the 27th Annual IEEE Conference on Local Computer Networks*, pages 333–342, November 2002.

[21] Chadi Barakat, Eitan Altman, and Walid Dabbous. On TCP performance in a heterogeneous network: a survey. *IEEE Communications Magazine*, 38(1):40–46, 2000.

[22] David Barkai. *Peer-to-Peer Computing: Technologies for Sharing and Collaborating on the Net*. Intel Press, Hillsboro, OR, 2001.

[23] Cynthia Barnhart and Yosef Sheffi. A network-based primal-dual heuristic for the solution of multicommodity network flow problems. *Transportation Science*, 27(2):102–117, 1993.

[24] Ivan D. Barrera, Gonzalo R. Arce, and Stephan Bohacek. Statistical approach for congestion control in gateway routers. *Computer Networks*, 55(3):572–582, 2011.

[25] Luis Barreto and Susana Sargento. TCP, XCP and RCP in wireless mesh networks: An evaluation study. In *Proceedings of the IEEE Symposium on Computers and Communications*, pages 351–357, June 2010.

[26] Mokhtar S. Bazaraa, Hanif D. Sherali, and C. M. Shetty. *Nonlinear Programming: Theory and Algorithms*. John Wiley and Sons, New York, 1993.

[27] Charalampos P. Bechlioulis and George A. Rovithakis. Robust adaptive control of feedback linearizable mimo nonlinear systems with prescribed performance. *IEEE Transactions on Automatic Control*, 53(9):2090–2099, 2008.

[28] Jon Bennett, Craig Partridge, and Nicholas Schectman. Packet reordering is not pathological network behavior. *IEEE/ACM Transactions on Networking*, 7(6):789–798, 1999.

[29] Alex Berson. *Client/Server Architecture*. McGraw-Hill, New York, 1992.

[30] Dimitri P. Bertsekas and Robert G. Gallager. *Data Networks 2nd Edition*. Prentice Hall, Englewood Cliffs, NJ, 1992.

[31] Wendemagegnehu T. Beyene and Jose E. Schutt-Aine. Efficient transient simulation of high-speed interconnects characterized by sampled data. *IEEE Transactions on Components, Packaging, and Manufacturing Technology, Part B: Advanced Packaging*, 21(1):105–114, February 1998.

[32] Shankar P. Bhattacharyya, Herve Chapellat, and Lee H. Keel. *Robust Control: The Parametric Approach 1st Edition*. Prentice Hall PTR, Upper Saddle River, NJ, 1995.

[33] Nooshin Bigdeli and Mohammad Haeri. AQM controller design for networks supporting TCP vegas: a control theoretical approach. *ISA Transactions*, 47(1):143–155, 2008.

[34] Christopher Bishop. *Neural Networks for Pattern Recognition*. Oxford University Press, New York, 1995.

[35] Anthony Bonato. *A Course on the Web Graph (Graduate Studies in Mathematics)*. American Mathematical Society, Boston, MA, 2008.

[36] Catherine Boutremans and Jean-Yves Le Boudec. A note on the fairness of TCP Vegas. In *Proceedings of the International Zurich Seminar on Broadband Communications*, pages 163–170, February 2000.

[37] Bob Braden, David Clark, and Jon Crowcroft. Recommendations on queue management and congestion avoidance in the internet. In *IETF RFC 2309*, April 1998.

[38] Scott Bradner. Key words for use in RFCs to indicate requirement levels. *RFC 2119*, March 1997.

[39] Lawrence Brakmo and Larry Peterson. TCP Vegas: end to end congestion avoidance on a global Internet. *IEEE Journal on Selected Areas in Communications*, 13(8):1465–1480, 1995.

[40] Lawrence S. Brakmo, Sean W. O'Malley, and Larry L. Peterson. TCP Vegas: new techniques for congestion detection and avoidance. In *Proceedings of the ACM Special Interest Group on Data Communication*, pages 24–35, October 1994.

[41] Torsten Braun, Michel Diaz, José Enriquez-Gabeiras, and Thomas Staub. *End-to-End Quality of Service over Heterogeneous Networks*. Springer-Verlag, Berlin, Heidelberg, Germany, 2008.

[42] Torsten Braun, Manuel Guenter, and Ibrahim Khalil. Management of quality of service enabled vpns. *IEEE Communications Magazine*, 39(5):90–98, 2001.

[43] David S. Broomhead and David Lowe. Multi-variable functional interpolation and adaptive networks. *Complex Systems 2*, pages 321–355, 1988.

[44] Hee-Jung Byun and Jong-Tae Lim. On a fair congestion control scheme for TCP Vegas. *IEEE Communications Letters*, 9(2):190–192, February 2005.

[45] Ramón Cáceres and Liviu Iftode. Improving the performance of reliable transport protocols in mobile computing environments. *IEEE Journal on Selected Areas in Communications*, 13(5):850–857, 1995.

[46] Michael J. Callaghan, Jim Harkin, E. McColgan, T. Martin McGinnity, and Liam P. Maguire. Client-server architecture for collaborative remote experimentation. *Journal of Network and Computer Applications*, 30(4):1295–1308, 2007.

[47] Claudio Casetti, Mario Gerla, Saverio Mascolo, Medy Y. Sanadidi, and Ren Wang. TCP Westwood: end-to-end congestion control for wired/wireless networks. *Wireless Networks*, 8(5):467–479, September 2002.

[48] Kartikeya Chandrayaan and Shiv Kalyanaraman. On impact of non-conformant flows on a network of droptail gateways. In *Proceedings of the IEEE Global Telecommunications Conference*, volume 6, pages 3123–3127, December 2003.

[49] George Chang, Marcus Healey, James A. M. McHugh, and T.L. Wang. *Mining the World Wide Web - An Information Search Approach (The Kluwer International Series on Information Retrieval, Volume 10) (The Information Retrieval Series)*. Springer, Boston, MA, 2001.

[50] Fu-Chuang Chen and Chen-Chung Liu. Adaptively controlling nonlinear continuous-time systems using multilayer neural networks. *IEEE Transactions on Automatic Control*, 39(6):1306–1310, 1994.

[51] Jiwei Chen, Fernando Paganini, M.Y. Sanadidi, Ren Wang, and Mario Gerla. Fluid-flow analysis of TCP Westwood with RED. *Computer Networks*, 50(9):1302–1326, 2006.

[52] Minghua Chen and Avideh Zakhor. Multiple tfrc connections based rate control for wireless networks. *IEEE Transactions on Multimedia*, 8(5):1045, 2006.

[53] Shigang Chen and Klara Nahrstedt. An overview of quality of service routing for next-generation high-speed networks: problems and solutions. *IEEE Network*, 12(6):64–79, 1998.

[54] Chang-Kuo Chena, Teh-Lu Liao, and Jun-Juh Yan. Active queue management controller design for TCP communication networks: Variable structure control approach. *Chaos, Solitons & Fractals*, 40(1):277–285, 2009.

[55] Rung-Shiang Cheng and Hui-Tang Lin. Protecting tcp from a misbehaving receiver. *International Journal of Network Management*, 17(3):209–218, June 2007.

[56] Dah-Ming Chiu and Raj Jain. Analysis of the increase and decrease algorithms for congestion avoidance in computer networks. *Journal Computer Networks and ISDN Systems*, 17(1):1–14, 1989.

[57] Joon-Young Choi, Kyungmo Koo, Jin S. Lee, and Steven H. Low. Global stability of FAST TCP in single-link single-source network. In *Proceedings of the 44th IEEE Conference on Decision and Control and European Control Conference*, December 2005.

[58] Joon-Young Choi, Kyungmo Koo, David X. Wei, Jin S. Lee, and Steven H. Low. Global exponential stability of FAST TCP. In *Proceedings of the 45th IEEE Conference on Decision and Control*, December 2006.

[59] Chrysostomos Chrysostomou, Andeas Pitsillides, and Y. Ahmet Sekercioglu. Fuzzy explicit marking: a unified congestion controller for Best-Effort and Diff-Serv networks. *Computer Networks*, 53(5):650–667, 2009.

[60] Daniel Coffman, Danny Soroker, Chandra Narayanaswami, and Aaron Zinman. A client-server architecture for state-dependent dynamic visualizations on the web. In *Proceedings of the 19th ACM International Conference on World Wide Web*, pages 1237–1240, July 2010.

[61] Michael A. Cohen and Stephen Grossberg. Absolute stability of global pattern formation and parallel memory storage by competitive neural networks. *IEEE Transactions on Systems, Man, and Cybernetics*, 13(5):815–826, 1983.

[62] Douglas E. Comer. *The Internet Book: Everything You Need to Know about Computer Networking and How the Internet Works*. Prentice-Hall, Inc., Upper Saddle River, NJ, 1995.

[63] Neil Cotter. The Stone-Weierstrass theorem and its application to neural networks. *IEEE Transactions on Neural Networks*, 1(4):290–295, 1990.

[64] George Cybenco. Approximation by superposition of a sigmoidal function. *Mathematics of Control, Signals and Systems*, 5(4):303–314, December 1989.

[65] Luiz A. DaSilva. QoS mapping along the protocol stack: discussion and preliminary results. In *Proceedings of the IEEE International Conference on Communications*, volume 2, pages 713–717, June 2000.

[66] Supratim Deb and Rayadurgam Srikant. Global stability of congestion controllers for the Internet. *IEEE Transactions on Automatic Control*, 48(6):1055–1060, 2003.

[67] Amir Dembo, Oluseyi Farotimi, and Thomas Kailath. High-order absolutely stable neural networks. *IEEE Transactions on Circuits and Systems*, 38(1):57–65, 1991.

[68] Vikas Deora, Jianhua Shao, W. Alex Gray, and Nick J. Fiddian. A quality of service management framework based on user expectations. *Lecture Notes in Computer Science*, 2910:104–114, 2003.

[69] Nikhil R. Devanur, Christos H. Papadimitriou, Amin Saberi, and Vijay V. Vazirani. Market equilibrium via a primal-dual algorithm for a convex program. *Journal of the ACM*, 55(5):1–18, November 2008.

[70] Dawei Ding, Jie Zhu, Xiaoshu Luo, and Yuliang Liu. Delay induced hopf bifurcation in a dual model of internet congestion control algorithm. *Nonlinear Analysis: Real World Applications*, 10(5):2873–2883, October 2009.

[71] Nandita Dukkipati, Tiziana Refice, Yuchung Cheng, Jerry Chu, Tom Herbert, Amit Agarwal, Arvind Jain, and Natalia Sutin. An argument

for increasing TCP's initial congestion window. *ACM SIGCOMM Computer Communication Review*, 40(3):26–33, June 2010.

[72] Nasif Ekiz, Abuthahir Habeeb Rahman, and Paul D. Amer. Misbehaviors in tcp sack generation. *ACM SIGCOMM Computer Communication Review*, 41(2):16–23, April 2011.

[73] Georgios John Fakas and Bill Karakostas. A peer to peer (p2p) architecture for dynamic workflow management. *Information and Software Technology*, 46(6):423–431, 2004.

[74] Jay A. Farrell and Marios M. Polycarpou. *Adaptive Approximation Based Control: Unifying Neural, Fuzzy and Traditional Adaptive Approximation Approaches (Adaptive and Learning Systems for Signal Processing, Communications and Control Series)*. Wiley-Interscience, New Jersey, 2006.

[75] Wu Chang Feng, Kang G. Shin, Dilip D. Kandlur, and Debanjan Saha. The BLUE active queue management algorithms. *IEEE/ACM Transactions on Networking*, 10(4):513–528, 2002.

[76] Victor Firoiu and Marty Borden. A Study of Active Queue Management for Congestion Control. In *Proceedings of the IEEE International Conference on Computer Communications 9th Annual Joint Conference of the IEEE Computer and Communications Societies*, volume 3, pages 1435–1444, March 2000.

[77] Paris Flegkas, Panos Trimintzios, and George Pavlou. A policy-based quality of service management system for ip diffserv networks. *IEEE Network*, 16(2):50–56, 2002.

[78] Sally Floyd. Connections with multiple congested gateways in packet-switched networks part 1: one-way traffic. *ACM Special Interest Group on Data Communication - Computer Communication Review*, 21(5):30–47, 1991.

[79] Sally Floyd. Metrics for the evaluation of congestion control mechanisms. *RFC 5166*, March 2008.

[80] Sally Floyd, Ramakrishna Gummadi, and Scott Shenker. Adaptive RED: An Algorithm for Increasing the Robustness of RED's Active Queue Management. *Technical Report*, 2001.

[81] Sally Floyd, Mark Handley, Jitendra Padhye, and Joerg Widmer. Equation based congestion control for unicast applications. In *Proceedings of the ACM SIGCOMM Conference on Applications, Technologies, Architectures, and Protocols for Computer Communication*, pages 43–56, August 2000.

[82] Sally Floyd and Tom Henderson. The NewReno modification to TCP's fast recovery algorithm. *RFC 2582*, 1999.

[83] Sally Floyd, Tom Henderson, and Andrei Gurtov. The NewReno modification to TCP's fast recovery algorithm. In *End-2-End-Interest Mailing List, RFC 3782*, 2004.

[84] Sally Floyd and Van Jacobson. Random early detection gateways for congestion avoidance. *IEEE/ACM Transactions on Networking*, 1(4):397–413, 1993.

[85] Chad Fogg, Didier J. LeGall, Joan L. Mitchell, and William B. Pennebaker. *MPEG Video Compression Standard (Digital Multimedia Standards Series)*. Kluwer Academic Publishers, New York, 1996.

[86] Jason A. S. Freeman and David Saad. Learning and generalization in radial basis function networks. *Neural Computation*, 7(5):1000–1020, 1995.

[87] Cheng P. Fu and Soung C. Liew. A remedy for performance degradation of TCP Vegas in asymmetric networks. *IEEE Communications Letters*, 7(1):42–44, 2003.

[88] Cheng Peng Fu, Ling-Chi Chung, and Soung C. Liew. Performance degradation of TCP Vegas in asymmetric networks and its remedies. In *Proceedings of the IEEE International Conference on Communications*, volume 10, pages 3229–3236, June 2001.

[89] Cheng Peng Fu, Associate Member, Soung C. Liew, and Senior Member. TCP Veno: TCP enhancement for transmission over wireless access networks. *IEEE Journal on Selected Areas in Communications*, 21(2):216–228, 2003.

[90] Kenji Funahashi. On the approximate realization of continuous mappings by neural networks. *Neural Networks*, 2(3):183–192, May 1989.

[91] Zhao Fuzhea, Zhou Jianzhonga, Luo Zhimenga, and Xiao Yang. Stability condition of FAST TCP in high speed network on the basis of control theory. *Journal of Systems Engineering and Electronics*, 19(4):843–850, 2008.

[92] Huijun Gao, James Lam, Changhong Wang, and Xinping Guan. Further results on local stability of REM algorithm with time-varying delays. *IEEE Communications Letters*, 9(5):402–404, 2005.

[93] Fei Ge, Liansheng Tan, and Moshe Zukerman. Throughput of FAST TCP in Asymmetric Networks. *IEEE Communications Letters*, 12(2):158–160, 2008.

[94] Shuzhi Sam Ge, Chang Chieh Hang, Tong Heng Lee, and Tao Zhang. *Stable Adaptive Neural Network Control 1st Edition*. Springer Publishing Company, New York, 2010.

[95] László Gerencsér. Closed loop parameter identifiability and adaptive control of a linear stochastic system. *Systems & Control Letters*, 15(5):411–416, 1991.

[96] Mario Gerla, Bryan K. F. Ng, M. Y. Sanadidi, Massimo Valla, and Ren Wang. TCP Westwood with adaptive bandwidth estimation to improve efficiency/friendliness tradeoffs. *Computer Communications*, 27(1):41–58, 2004.

[97] Mario Gerla, Medy Y. Sanadidi, Ren Wang, and Andrea Zanella. TCP Westwood: congestion window control using bandwidth estimation. In *Proceedings of the IEEE Global Telecommunications Conference*, volume 3, pages 1698–1702, November 2001.

[98] Richard Gibbens and Frank Kelly. Resource pricing and the evolution of congestion control. *Automatica*, 35(12):1969–1985, 1999.

[99] Anna C. Gilbert, Youngmi Joo, and Nick Mckeown. Congestion control and periodic behavior. In *Proceedings of the 11th IEEE Workshop on Local and Metropolitan Area Networks*, March 2001.

[100] Daniel Gmach, Stefan Krompass, Andreas Scholz, Martin Wimmer, and Alfons Kemper. Adaptive quality of service management for enterprise services. *ACM Transactions on the Web*, 2(1):1–46, 2008.

[101] Jamaloddin Golestani and Krishan Sabnani. Fundamental observations on multicast congestion control in the Internet. In *Proceedings of IEEE International Conference on Computer Communications 8th Annual Joint Conference of the IEEE Computer and Communications Societies*, March 1999.

[102] Hiroaki Gomi and Mitsuo Kawato. Neural network control for a closed-loop system using feedback-error-learning. *Neural Network*, 6(7):933–946, 1993.

[103] G. C. Goodwin and K. S. Sin. *Adaptive Filtering: Prediction and Control*. Prentice Hall, New Jersey, 1984.

[104] Roch Guerin, Hamid Ahmadi, and Mahmoud Naghshineh. Equivalent capacity and its application to bandwidth allocation in high-speed networks. *IEEE Journal on Selected Areas in Communications*, 9(7):968–981, 1991.

[105] Roch Guerin and Vinod G. J. Peris. Quality-of-service in packet networks: basic mechanisms and directions. *Computer Networks*, 31(3):169–189, 1999.

[106] Andre Gunther and Christian Hoene. Measuring round trip times to determine the distance between wlan nodes. In *Proceedings of the Networking '05. Networking Technologies, Services, and Protocols; Performance of Computer and Communication Networks; Mobile and Wireless Communications Systems*, volume 3462, pages 303–319, May 2005.

[107] Madan Gupta and Dandina Rao. *Neuro Control Systems: Theory and Applications*. IEEE Press, New York, 1993.

[108] Andrei Gurtov. Making TCP robust against spikes. In *Technical Report C-2001-53*, 2001.

[109] Andrei Gurtov. Effect of delays on tcp performance. In *Proceedings of the IFIP TC6/WG6.8 Working Conference on Emerging Personal Wireless Communications*, volume 67, pages 87–105, 2002.

[110] Andrei Gurtov and Sally Floyd. Modeling wireless links for transport protocols. *ACM Computer Communications Review*, 34(2):85–96, April 2004.

[111] Abdelhakim Hafid and Gregor V. Bochmann. An approach to quality of service management in distributed multimedia application: design and an implementation. *Multimedia Tools and Applications*, 9(2):167–191, 1999.

[112] Fred Halsall. *Data Communications, Computer Networks and Open Systems*. Addison-Wesley, London, United Kingdom, 1996.

[113] Huaizhong Han, Christopher V. Hollot, Donald F. Towsley, and Yossi Chait. Synchronization of TCP flows in networks with small drop tail buffers. In *Proceedings of the 44th IEEE Conference on Decision and Control and 2005 European Control Conference*, pages 6762–6767, December 2005.

[114] Go Hasegawa, Kenji Kurata, and Masayuki Murata. Analysis and improvement of fairness between TCP Reno and Vegas for deployment of TCP Vegas to the Internet. In *Proceedings of the IEEE International Conference on Network Protocols*, pages 177–186, November 2000.

[115] Mahbub Hassan and Raj Jain. *High Performance TCP/IP Networking:Concepts, Issues, and Solutions*. Prentice Hall, New Jersey, 2004.

[116] Simon Hauger, Michael Scharf, Jochen Kogel, and Chawapong Suriyajan. Evaluation of router implementations for explicit congestion control schemes. *Journal of Communications*, 5(3):197–204, March 2010.

[117] Howard Paul Hayden. Voice flow control in integrated packet networks. In *Technical Report MIT/LIDS/TH/TR-601, Laboratory for Information and Decision Systems, Massachusetts Institute of Technology*, 1981.

[118] Simon Haykin. *Neural Networks: A Comprehensive Foundation 2nd Edition*. Prentice Hall, New York, 1995.

[119] Donald W. Hearn and Siriphong Lawphongpanich. A dual ascent algorithm for traffic assignment problems. *Transportation Research Part B: Methodological*, 24(6):423–430, 1990.

[120] Zhang Heying, Liu Baohong, and Dou Wenhua. Design of a robust active queue management algorithm based on feedback compensation. In *Proceedings of ACM Special Interest Group on Data Communication*, pages 277–285, August 2003.

[121] Fu-Sheng Ho and Petros A. Ioannou. Traffic flow modeling and control using artificial neural networks. *IEEE Control Systems*, 16(5):16–26, 1996.

[122] Markus Hofmann and Leland R. Beaumont. *Content Networking: Architecture, Protocols, and Practice (The Morgan Kaufmann Series in Networking)*. Morgan Kaufmann Publishers, San Francisco, CA, 2005.

[123] Christopher V. Hollot, Vishal Misra, Don Towsley, and Weibo Gong. Analysis and design of controllers for AQM routers supporting TCP flows. *IEEE Transactions on Automatic Control*, 47(6):945–959, June 2002.

[124] John Joseph Hopfield. Neurons with graded response have collective computational properties like those of two-state neurons. In *Proceedings of the National Academy of Sciences*, volume 81, pages 3088–3092, May 1984.

[125] Christos N. Houmkozlis and George A. Rovithakis. A neural network congestion control algorithm for the internet. In *Proceedings of the 2005 IEEE International Symposium on Intelligent Control, Mediterrean Conference on Control and Automation*, pages 450–455, June 2005.

[126] Christos N. Houmkozlis and George A. Rovithakis. Towards satisfying an application level quality measure via a neural network tcp-like protocol. In *Proceedings of the 44th IEEE Conference on Decision and Control and the European Control Conference*, pages 6946–6951, December 2005.

[127] Christos N. Houmkozlis and George A. Rovithakis. A neuro-adaptive TCP-like protocol with cost constraints. In *Proceedings of the 14th Mediterranean Conference on Control and Automation*, pages 1–6, June 2006.

[128] Christos N. Houmkozlis and George A. Rovithakis. A neuro-adaptive TCP-like protocol with communication channels adaptation under cost constraints. In *Proceedings of the European Control Conference*, July 2007.

[129] Christos N. Houmkozlis and George A. Rovithakis. A robust neuro-adaptive congestion control scheme with respect to exogenous disturbances and delay. In *Proceedings of the 15th Mediterranean Conference on Control and Automation*, pages 1–6, June 2007.

[130] Christos N. Houmkozlis and George A. Rovithakis. A neuro-adaptive congestion control scheme for round trip regulation. *Automatica*, 44(5):1402–1410, 2008.

[131] Christos N. Houmkozlis and George A. Rovithakis. Fairness guarantees in a neural network adaptive congestion control framework. *IEEE Transactions on Neural Networks*, 20(3):527–533, 2009.

[132] Xiaomeng Huang, Chuang Lin, and Fengyuan Ren. Generalized modeling and stability analysis of HighSpeed TCP and scalable TCP. *IEICE Transactions on Communications*, E89-B(2):605–608, 2006.

[133] Jean-Francois Huard and Aurel A. Lazar. On QoS mapping in multimedia networks. In *The Twenty-First Annual International Computer Software and Applications Conference*, pages 312–317, August 1997.

[134] Gianluca Iannaccone, Sharad Jaiswal, and Christophe Diot. Packet reordering inside the Sprint backbone. In *Sprint ATL Technical Report TR01-ATL-062917*, 2001.

[135] Hiroshi Inamura, Gabriel Montenegro, Reiner Ludwig, Andrei Gurtov, and Farid Khafizov. TCP over second (2.5G) and third (3G) generation wire. *Technical Report RFC 3481*, 2003.

[136] Petros A. Ioannou and Andreas Pitsillides. *Modeling and Control of Complex Systems*. CRC Press, Boca Raton, FL, 2008.

[137] Petros A. Ioannou and Jing Sun. *Robust Adaptive Control*. Prentice Hall, Los Angeles, CA, 1996.

[138] Yoshihiro Ito and Shuji Tasaka. Quantitative assessment of user-level qos and its mapping. *IEEE Transactions on Multimedia*, 7(3):572–584, 2005.

[139] Van Jacobson. Congestion avoidance and control. In *Proceedings of ACM Special Interest Group on Data Communication - Symposium Proceedings on Communications Architectures and Protocols*, November 1988.

[140] Van Jacobson. Berkeley TCP evolution from 4.3-Tahoe to 4.3-Reno. In *Proceedings of the 18th Internet Engineering Task Force*, pages 365–376, August 1990.

[141] Van Jacobson. Modified TCP congestion control avoidance algorithm. In *End-2-End-Interest Mailing List*, April 1990.

[142] Van Jacobson and Michael J. Karels. Congestion avoidance and control. In *Proceedings of ACM Special Interest Group on Data Communication*, pages 314–329, August 1988.

[143] Krister Jacobsson, Lachlan L. H. Andrew, Ao Tang, Steven H. Low, and Håkan Hjalmarsson. An improved link model for window flow control and its application to FAST TCP. *IEEE Transactions on Automatic Control*, 54(3):551–564, 2009.

[144] Jeffrey M. Jaffe. A Decentralized Optimal Multiple-User Flow Control Algorithm. *IEEE Transaction on Communications*, pages 954–962, 1981.

[145] Suresh Jagannathan and Frank L. Lewis. Identification of nonlinear dynamical systems using multilayered neural networks. *Automatica*, 32(12):1707–1712, 1996.

[146] Raj Jain. *The Art of Computer Systems Performance Analysis: Techniques for Experimental Design, Measurement, Simulation and Modeling*. John Wiley and Sons, New York, 1991.

[147] Sharad Jaiswal, Gianluca Iannaccone, Christophe Diot, Jim Kurose, and Don Towsley. Measurement and classification of out-of-sequence packets in a tier-1 IP backbone. volume 15, pages 54–66, 2007.

[148] Taher M. Jelleli and Adel M. Alimi. On the applicability of the minimal configured hierarchical fuzzy control and its relevance to function approximation. *Applied Soft Computing*, 9(4):1273–1284, September 2009.

[149] Hao Jiang and Constantinos Dovrolis. Passive estimation of tcp round-trip times. *ACM SIGCOMM Computer Communication Review*, 32(3):75–88, 2002.

[150] Cheng Jin, David X. Wei, and Steven H. Low. FAST TCP: motivation, architecture, algorithms, performance. In *Proceedings of IEEE International Conference on Computer Communications Twenty-Third Annual Joint Conference of the IEEE Computer and Communications Societies*, pages 2490–2501, March 2004.

[151] Cheng Jin, David X. Wei, Steven H. Low, Julian J. Bunn, Hyojeong D. Choe, John C.Doylle, Harvey B. Newman, Sylvain Ravot, Suresh Singh, Fernando Paganini, Gary Buhrmaster, Roger Les Cottrell, Olivier Martin, and Wu chun Feng. FAST TCP: from theory to experiments. *IEEE Network*, 19(1):4–11, 2005.

[152] Tian Jin, Xiangzhi Sheng, and Wenjun Wu. The effect on the interfairness of TCP and TFRC by the phase of TCP traffics. In *International Conference on Computer Networks and Mobile Computing*, pages 131–136, October 2001.

[153] Ramesh Johari and David Kim Hong Tan. End-to-end congestion control for the Internet: delays and stability. *IEEE/ACM Transactions on Networking*, 9(6):818–832, 2001.

[154] Manali Joshi, Ajay Mansata, Salil Talauliker, and Cory Beard. Design and analysis of multi-level active queue management mechanisms for emergency traffic. *Computer Communications*, 28(2):162–173, February 2005.

[155] Yves Kamp and Martin Hasler. *Recursive Neural Networks for Associative Memory*. John Wiley and Sons, New York, 1990.

[156] Koushik Kar, Saswati Sarkar, and Leandros Tassiulas. A primal algorithm for optimization based rate control for unicast sessions. In *TR 2000-22*.

[157] Phil Karn and Craig Partridge. Improving round-trip time estimates in reliable transport protocols. In *Proceedings of the ACM Workshop on Frontiers in Computer Communications Technology*, pages 2–7, November 1987.

[158] Dina Katabi and Charles Blake. A note on the stability requirements of adaptive virtual queue. *MIT Technical Memo*, 2002.

[159] Frank P. Kelly, Aman K. Maulloo, and David K. H. Tan. Rate control for communication networks: shadow prices, proportional fairness and stability. *Journal of the Operational Research Society*, 49(3):237–252, 1998.

[160] Tom Kelly. Scalable TCP: improving performance in highspeed wide area networks. In *Proceedings of the ACM Special Interest Group on Data Communication Computer Communication*, pages 83–91, April 2003.

[161] Hassan Khalil. *Nonlinear Systems*. Prentice Hall, New Jersey, 2002.

[162] Dongkeun Kim and Jaiyong Lee. End-to-end one-way delay estimation using one-way delay variation and round-trip time. In *Proceedings of the Fourth International Conference on Heterogeneous Networking for Quality, Reliability, Security and Robustness*, pages 19:1–19:8, August 2007.

[163] Ki Baek Kim. Design of feedback controls supporting TCP based on the state space approach. *IEEE Transactions on Automatic Control*, 51(7):1086–1099, 2006.

[164] Tae-Hoon Kim and Kee-Hyun Lee. Refined adaptive RED in TCP/IP networks. In *Proceedings of SICE-ICASE International Joint Conference*, pages 3722–3725, October 2006.

[165] Csaba Király, Michele Garetto, Michela Meo, Marco Ajmone Marsan, and Renato Lo Cigno. Analytical computation of completion time distributions of short-lived tcp connections. *Performance Evaluation*, 59(2-3):179–197, 2005.

[166] Tokuda Koichi, Hasegawa Go, and Murata Masayuki. Performance analysis of highspeed tcp and its improvement for high throughput and fairness against tcp reno connections. *Institute of Electronics, Information and Communication Engineers Technical Report*, 102(694):213–218, 2003.

[167] Kyungmo Koo, Joon-Young Choi, and Jin Soo Lee. Parameter conditions for global stability of FAST TCP. *IEEE Communications Letters*, 12(2):155–157, 2008.

[168] Elias B. Kosmatopoulos, Marios M. Polycarpou, Manolis A. Christodoulou, and Petros A. Ioannou. High-order neural network structures for identification of dynamical systems. *IEEE Transactions on Neural Networks*, 6(2):422–431, 1995.

[169] Charles M. Kozierok. *The TCP/IP Guide: A Comprehensive, Illustrated Internet Protocols Reference*. William Pollock, San Francisco, CA, 2005.

[170] Ryogo Kubo, Junichi Kani, and Yukihiro Fujimoto. Congestion control in TCP/AQM networks using a disturbance observer. *IEEJ Transactions on Industry Applications*, 129(6):541–547, 2009.

[171] Srisankar Kunniyur and Rayadurgam Srikant. Analysis and design of an adaptive virtual queue (AVQ) algorithm for active queue management. In *Proceedings of ACM Special Interest Group on Data Communication Conference on Applications, Technologies, Architectures, and Protocols For Computer Communications*, pages 123–134, August 2001.

[172] Jim Kurose. Open numbers and challenges in providing quality of service guarantees in high-speed networks. *ACM SIGCOMM - Computer Communication Review*, 23(1):6–15, 1993.

[173] Richard J. La, Priya Ranjan, and Eyad H. Abed. Analysis of adaptive random early detection (ARED). In *18th International Teletraffic Congress*, pages 53–67, 2003.

[174] T. V. Lakshman and Upamanyu Madhow. The performance of TCP/IP for networks with high bandwidth-delay products and random loss. *IEEE/ACM Transactions on Networking*, 5(3):336–350, 1997.

[175] Stephen Lane, David Handelman, and Jack Gelfand. Theory and development of higher-order cmac neural networks. *IEEE Control Systems*, 12(2):23–30, 1992.

[176] Michael Laor and Lior Gendel. The effect of packet reordering in a backbone link on application throughput. *IEEE Network*, 16(5):28–36, 2002.

[177] Nicolas Larrieu and Philippe Owezarski. TFRC contribution to internet QoS improvement. *Lecture Notes in Computer Science*, 2811:73–82, 2003.

[178] Torbjorn Larsson and Michael Patriksson. An augmented lagrangean dual algorithm for link capacity side constrained traffic assignment problems. *Transportation Research Part B: Methodological*, 29(6):433–455, 1995.

[179] Marios Lestas, Andreas Pitsillides, Petros A. Ioannou, and George Hadjipollas. Adaptive congestion protocol: a congestion control protocol with learning capability. *Computer Networks*, 51(13):3773–3798, September 2007.

[180] Ka-Cheong Leung, Victor O. K. Li, and Daiqin Yang. An overview of packet reordering in transmission control protocol (TCP): problems, solutions, and challenges. *IEEE Transactions on Parallel and Distributed Systems*, 18(4):522–535, 2007.

[181] Pingkang Li, Uwe Kruger, and George W. Irwin. Identification of dynamic systems under closed-loop control. *International Journal of Systems Science*, 37(3):181–195, 2006.

[182] Qi Li and Di Chen. Analysis and improvement of TFRC congestion control mechanism. In *Proceedings of the International Conference on Wireless Communications, Networking and Mobile Computing*, pages 1149–1153, September 2005.

[183] Yee-Ting Li, Douglas Leith, and Robert N. Shorten. Experimental evaluation of TCP protocols for high-speed networks. *IEEE/ACM Transactions on Networking*, 15(5):1109–1122, 2007.

[184] Dong Lin and Robert Morris. Dynamics of random early detection. In *Proceedings of the ACM Special Interest Group on Data Communication Conference on Applications, Technologies, Architectures, and Protocols for Computer Communication*, pages 127–137, October 1997.

[185] Cricket Liu, Adrian Nye, Jerry Peek, Bryan Buus, and Russ Jones. *Managing Internet Information Services: World Wide Web, Gopher, FTP, and More*. O'Reilly Media, Sebastopol, CA, 1994.

[186] Shao Liu, Tamer Başar, and Rayadurgam Srikant. Controlling the Internet: a survey and some new results. In *Proceedings of the 42nd IEEE Conference on Decision and Control*, volume 3, pages 3048–3057, December 2003.

[187] Shao Liu, Tamer Başar, and Rayadurgam Srikant. TCP-Illinois: a loss-and delay-based congestion control algorithm for high-speed networks. *Performance Evaluation*, 65(6-7):417–440, 2008.

[188] Yong Liu, Yang Guo, and Chao Liang. A survey on peer-to-peer video streaming systems. *Peer-to-Peer Networking and Applications*, 1(1):18–28, 2008.

[189] Cheng-Nian Long, Bin Zhao, and Xin-Ping Guan. SAVQ: stabilized adaptive virtual queue management algorithm. *IEEE Communications Letters*, 9(1):78–80, 2005.

[190] Mary E. S. Loomis. Client-server architecture. *Journal of Object Oriented Programming*, 4(6):40–44, 1992.

[191] Pete Loshin. *Big Book of IP Telephony RFCs*. Morgan Kaufmann Publishers, San Francisco, CA, 2001.

[192] Pete Loshin. *TCP/IP Clearly Explained*. Morgan Kaufmann Publishers, San Francisco, CA, 2002.

[193] S.H. Low and D.E. Lapsley. Optimization flow control I: basic algorithm and convergence. *IEEE/ACM Transactions on Networking*, 7(6):861–874, 1999.

[194] Steven H. Low. A duality model of TCP and queue management algorithms. *IEEE/ACM Transactions on Networking*, 11(4):525–536, August 2003.

[195] Steven H. Low, Lachlan L. H. Andrew, and Bartek P. Wydrowski. Understanding XCP: equilibrium and fairness. In *Proceedings of the IEEE International Conference on Computer Communications*, volume 2, pages 1025–1036, June 2005.

[196] Steven H. Low, Fernando Paganini, Jiantao Wang, Sachin Adlakha, and John C. Doyle. Dynamics of TCP/RED and a scalable control. In *Proceedings of IEEE International Conference on Computer Communications Twenty-First Annual Joint Conference of the IEEE Computer and Communications Societies*, June 2002.

[197] Steven H. Low, Larry Peterson, and Imin Wang. Understanding TCP Vegas: a duality model. In *Proceedings of ACM SIGMETRICS International Conference on Measurement and Modeling of Computer Systems*, pages 226–235, June 2001.

[198] Xiapu Luo, Rocky K. C. Chang, and Edmond W. W. Chan. Performance analysis of TCP/AQM under denial-of-service attacks. In *Proceedings of the 13th IEEE International Symposium on Modeling, Analysis, and Simulation of Computer and Telecommunication Systems*, pages 97–104, September 2005.

[199] Kexin Ma, Ravi R. Mazumdar, and Jun Luo. On the performance of primal-dual schemes for congestion control in networks with dynamic flows. In *Proceedings of 27th IEEE International Conference on Computer Communications*, pages 326–330, April 2008.

[200] Johann M. Hofmann Magalhaes and Paulo R. Guardieiro. A new qos mapping for streamed mpeg video over a diffserv domain. In *Proceedings of the IEEE International Conference on Communications, Circuits and Systems and West Sino Expositions*, volume 1, pages 675–679, June 2002.

[201] Jamshid Mahdavi and Sally Floyd. TCP-friendly unicast rate-based flow control. *End-2-End Interest Mailing List*, 1997.

[202] Richard Marquez, Eitan Altman, and Solazver Solé-Álvarez. Modeling TCP and high speed TCP: a nonlinear extension to AIMD mechanisms. *Lecture Notes in Computer Science*, 3079:132–143, 2004.

[203] Jim Martin, Arne Nilsson, and Injong Rhee. Delay-based congestion avoidance for TCP. *IEEE/ACM Transactions on Networking*, 11(3):356–369, 2003.

[204] Laurent Massoulie. Stability of distributed congestion control with heterogeneous feedback delays. *IEEE Transactions on Automatic Control*, 47(6):895–902, 2002.

[205] Martin May, Thomas Bonald, and Jean-Chrysostome Bolot. Analytic evaluation of RED performance. In *Proceedings of the IEEE International Conference on Computer Communications 9th Annual Joint Conference of the IEEE Computer and Communications Societies*, volume 3, pages 1415–1424, March 2000.

[206] Deepankar Medhi and Karthikeyan Ramasamy. *Network Routing: Algorithms, Protocols, and Architectures (The Morgan Kaufmann Series in Networking)*. Morgan Kaufmann Publishers, San Francisco, CA, 2007.

[207] Shashidhar Merugu, Sridhar Srinivasan, and Ellen W. Zegura. Adding structure to unstructured peer-to-peer networks: the use of small-world graphs. *Journal of Parallel and Distributed Computing*, 65(2):142–153, 2005.

[208] Michael Miller. *Cloud Computing: Web-Based Applications That Change the Way You Work and Collaborate Online*. Que, Indianapolis, IN, 2008.

[209] Jeonghoon Mo, Richard J. La, Venkat Anantharam, and Jean Walrand. Analysis and comparison of TCP Reno and Vegas. In *Proceedings of IEEE International Conference on Computer Communications Eighteenth Annual Joint Conference of the IEEE Computer and Communications Societies*, pages 1556–1563, March 1999.

[210] Jeonghoon Mo and Jean Walrand. Fair end-to-end window-based congestion control. *IEEE/ACM Transactions on Networking*, 8(5):556–567, October 2000.

[211] Bogdan Moraru, Flavius Copaciu, Gabriel Lazar, and Virgil Dobrota. Practical analysis of TCP implementations: Tahoe, Reno, NewReno. In *Proceedings of the 2nd RoEduNet International Conference*, pages 125–130, June 2003.

[212] Peter Morville and Louis Rosenfeld. *Information Architecture for the World Wide Web: Designing Large-Scale Web Sites*. O'Reilly Media, Sebastopol, CA, 2006.

[213] Masayoshi Nabeshima and Kouji Yata. Improving the convergence time highspeed TCP. In *Proceedings of 12th IEEE International Conference on Networks*, pages 19–23, November 2004.

[214] Dhinaharan Nagamalai, Seoung-Hyeon Lee, Won-Goo Lee, and Jae-Kwang Lee. SCTP over high speed wide area networks. *Lecture Notes in Computer Science*, 3420:628–634, 2005.

[215] Tatsuo Nakajima. Resource reservation for adaptive qos mapping in real-time mach. *Parallel and Distributed Processing, Lecture Notes in Computer Science*, 1388:1047–1056, 1998.

[216] Giovanni Neglia, Vincenzo Falletta, and Giuseppe Bianchi. Is TCP packet reordering always harmful? In *Proceedings of IEEE Computer Societies 12th Annual International Symposium on Modeling, Analysis, and Simulation of Computer and Telecommunications Systems*, pages 87–94, October 2004.

[217] Oliver Nelles. *Nonlinear System Identification*. Springer, Berlin, Germany, 2001.

[218] Natalia Olifer and Victor Olifer. *Computer Networks: Principles, Technologies and Protocols for Network Design*. John Wiley and Sons, New York, 2006.

[219] Carlos Oliveira, Jaime Bae Kim, and Tatsuya Suda. An adaptive bandwidth reservation scheme for high-speed multimedia wireless networks. *IEEE Journal on Selected Areas in Communications*, 16(6):858–874, 1998.

[220] Andy Oram. *Peer-to-Peer: Harnessing the Power of Disruptive Technologies*. O'Reilly & Associates, Sebastopol, CA, 2001.

[221] F. Paganini, Z. Wang, Z. J. Doyle, and S. Low. Congestion control for high performance, stability, and fairness in general networks. *IEEE/ACM Transactions on Networking*, 13(1):43–56, 2005.

[222] Fernando Paganini, Zhikui Wang, Steven Low, and John Doyle. A new TCP/AQM for stable operation in fast networks. In *Proceedings of IEEE International Conference on Computer Communications 22nd Annual Joint Conference of the IEEE Computer and Communications Societies*, pages 96–105, March 2003.

[223] Rong Pan, Balaji Prabhakar, and Konstantinos Psounis. Choke - a stateless active queue management scheme for approximating fair bandwidth allocation. In *Proceedings of IEEE International Conference on Computer Communications 9th Annual Joint Conference of the IEEE Computer and Communications Societies*, pages 942–951, March 2000.

[224] Qixiang Pang, Soung C. Liew, Cheng Peng Fu, Wei Wang, and Victor O. K. Li. Performance study of TCP Veno over WLAN and RED router. In *Proceedings of the IEEE Global Telecommunications Conference*, pages 447–451, December 2003.

[225] Vern Paxon. End-to-end internet packet dynamics. *IEEE/ACM Transactions on Networking*, 7(3):277–292, 1999.

[226] Pierre Peretto and Jean-Jacques Niez. Long term memory storage capacity of multiconnected neural networks. *Biological Cybernetics*, 54(1):53–63, 1986.

[227] Damien Phillips and Jiankun Hu. Analytic models for highspeed TCP fairness analysis. In *Proceedings of 12th IEEE International Conference on Networks*, pages 725–730, November 2004.

[228] Nischal Piratla, Anura Jayasumana, and Abhijit Bare. Reorder density (RD): a formal, comprehensive metric for packet reordering. In *Proceedings of International Federation for Information Processing Networking Conference*, pages 78–89, May 2005.

[229] Tomaso Poggio and Federico Girosi. Regularization algorithms for learning that are equivalent to multilayer networks. *Science*, 247(4945):978–982, 1990.

[230] Marios M. Polycarpou. Stable adaptive neural control scheme for nonlinear systems. *IEEE Transactions on Automatic Control*, 41(3):447–451, 1996.

[231] Marios M. Polycarpou and Petros A. Ioannou. A robust adaptive nonlinear control design. *Automatica*, 32(3):423–427, 1996.

[232] M.J.D. Powell. Radial basis functions for multivariable interpolation: a review. In *Proceedings of the IMA Conference on Algorithms for the Approximation of Functions and Data*, pages 143–167, 1985.

[233] Lili Qiu, Yin Zhang, and Srinivasan Keshav. On individual and aggregate TCP performance. In *Proceedings of Seventh International Conference on Network Protocols*, pages 203–212, November 1999.

[234] Bozidar Radunovic and Jean-Yves Le Boudec. A unified framework for max-min and min-max fairness with applications. *IEEE/ACM Transactions on Networking*, 15(5):1073–1083, 2007.

[235] Gaurav Raina. Local bifurcation analysis of some dual congestion control algorithms. *IEEE Transactions on Automatic Control*, 50(8):1135–1146, 2005.

[236] Satish Rege. A distributed system client/server architecture for interactive multimedia applications. In *Proceedings of the 41st IEEE International Computer Conference*, pages 211–218, 1996.

[237] Jacqueline R. Reich. Design and implementation of a client-server architecture for taxonomy manager. *Software - Practice and Experience*, 29(2):143–166, 1999.

[238] Reza Rejaie, Mark Handley, and Deborah Estrin. RAP: An end-to-end rate-based congestion control mechanism for realtime streams in the internet. In *Proceedings of IEEE International Conference on Computer Communications Eighteenth Annual Joint Conference of the IEEE Computer and Communications Societies*, pages 1337–1345, 1999.

[239] Injong Rhee and Lisong Xu. CUBIC: a new TCP-friendly high-speed TCP variant. In *Proceedings of International Workshop on Protocols for Fast Long-Distance Networks*, pages 64–74, February 2005.

[240] John Rhoton. *Cloud Computing Explained: Implementation Handbook for Enterprises*. Recursive Press, London, United Kingdom, 2009.

[241] John Ross. *Discover the World Wide Web (Six-Point Discover Advantage)*. John Wiley and Sons, New York, 1997.

[242] Matthew Roughan, Ashok Erramilli, and Darryl Veitch. Network performance for TCP networks part i: persistent sources. In *Proceedings of the 17th International Teletraffic Congress–ITC-17*, pages 857–868, September 2001.

[243] George A. Rovithakis and Manolis A. Christodoulou. Adaptive control of unknown plants using dynamical neural networks. *IEEE Transactions on Systems, Man and Cybernetics*, 24(3):400–412, 1994.

[244] George A. Rovithakis and Manoulis A. Christodoulou. *Adaptive Control with Recurrent High-Order Neural Networks: Theory and Industrial Applications*. Springer, London, United Kingdom, 2000.

[245] George A. Rovithakis, Athanasios G. Malamos, Theodora Varvarigou, and Manolis A. Christodoulou. Quality assurance in networks - a high order neural net approach. In *Proceedings of Decision and Control Conference*, pages 1599–1604, December 1998.

[246] George A. Rovithakis, Athanasios G. Malamos, Theodora Varvarigou, and Manolis A. Christodoulou. Quality of service adaptive control in multimedia telecommunication services. In *Reliability, Survivability and Quality of Large Scale Telecommunication Systems*, pages 120–141. John Wiley and Sons, 2003.

[247] Goerge A. Rovithakis. Stable adaptive neuro-control design via Lyapunov function derivative estimation. *Automatica*, 37(8):1213–1221, 2001.

[248] David E. Rumelhart and James L. McClelland. *Parallel Distributed Processing: Explorations in the Microstructure of Cognition*. MIT Press, Cambridge, MA, 1986.

[249] Nader Sadegh. A perceptron network for functional identification and control of nonlinear systems. *IEEE Transactions on Neural Networks*, 4(6):982–988, 1993.

[250] Yusuke Sakumoto, Hiroyuki Ohsaki, and Makoto Imase. On XCP stability in a heterogeneous network. In *Proceedings of 12th IEEE Symposium on Computers and Communications*, July 2007.

[251] Charalampos (Babis) Samios and Mary K. Vernon. Modeling the throughput of TCP Vegas. In *Proceedings of ACM SIGMETRICS International Conference on Measurement and Modeling of Computer Systems*, pages 71–81, June 2003.

[252] Elisa Schae. Review of a course on the web graph by Anthony Bonato (American Mathematical Society, Providence, RI. *SIGACT News*, 40(2):35–37, June 2009.

[253] Henning Schulzrinne, Jim Kurose, and Don Towsley. Congestion control for real-time traffic in high-speed networks. In *Proceedings of the Ninth Annual Joint Conference of the IEEE Computer and Communication Societies. The Multiple Facets of Integration.*, volume 2, pages 543–550, June 1990.

[254] Rich Seifert and James Edwards. *The All-New Switch Book: The Complete Guide to LAN Switching Technology*. Wiley Publishing, Indianapolis, IN, 2008.

[255] Eric Setton and Bernd Girod. *Peer-to-Peer Video Streaming*. Springer Science Business Media, New York, 2007.

[256] Vinod Sharma, Jorma T. Virtamo, and Pasi E. Lassila. Performance analysis of the random early detection algorithm. *Probability in the Engineering and Informational Sciences*, 16(3):367–388, 2002.

[257] Hong Shen, Lin Cai, and Xuemin Shen. Performance analysis of TFRC over wireless link with truncated link-level ARQ. *IEEE Transactions on Wireless Communications*, 5(6):1479–1487, 2006.

[258] Jitae Shin, Jin-Gyeong Kim, Jongwon Kim, and C. C. Jay Kuo. Dynamic qos mapping control for streaming video in relative service differentiation networks. *European Transactions on Telecommunications*, 12(3):217–229, 2001.

[259] Balázs Sonkoly, Tuan Anh Trinh, and Sándor Molnár. Understanding highspeed TCP: a control-theoretic perspective. In *Proceedings of Communications and Computer Networks*, pages 74–80, October 2005.

[260] Barrie Sosinsky. *Cloud Computing Bible*. John Wiley and Sons, New York, 2011.

[261] Jeffrey T. Spooner, Raul Ordonez, Manfredi Maggiore, and Kevin M. Passino. *Stable Adaptive Control and Estimation for Nonlinear Systems: Neural and Fuzzy Approximation Techniques*. John Wiley and Sons, New York, NY, 2001.

[262] Rayadurgam Srikant. *The Mathematics of Internet Congestion Control*. Birkhäuser, Boston, MA, 2004.

[263] William Stallings. *Local Networks: An Introduction*. Prentice Hall PTR, Upper Saddle River, NJ, 1984.

[264] Richard Stevens. TCP slow start, congestion avoidance, fast retransmit, and fast recovery algorithms. *RFC 2001*, January 1997.

[265] Alexander L. Stolyar. Maximizing queueing network utility subject to stability: greedy primal-dual algorithm. *Queueing Systems*, 50(4):401–457, 2005.

[266] Alexander L. Stolyar. Greedy primal-dual algorithm for dynamic resource allocation in complex networks. *Queueing Systems*, 54(3):203–220, 2006.

[267] Tim Szigeti and Christina Hattingh. *End-to-end QoS network design: quality of service in LANs, WANs, and VPNs*. Cisco Press, Indianapolis, IN, 2008.

[268] Keiichi Takagaki, Hiroyuki Ohsaki, and Masayuki Murata. Analysis of a window-based flow control mechanism based on TCP Vegas in heterogeneous network environment. In *Proceedings of IEEE International Conference on Communications*, pages 1603–1606, June 2001.

[269] Liansheng Tan, Gang Peng, and Sammy Chan. Adaptive REM: random exponential marking with improved robustness. *Electronics Letters*, 43(2):133–135, 2007.

[270] Liansheng Tan, Wei Zhang, and Cao Yuan. On parameter tuning for fast tcp. *IEEE Communications Letters*, 11(5):458–460, 2005.

[271] Andrew S. Tanenbaum. *Computer Networks, Fourth Edition*. Prentice Hall, Upper Saddle River, NJ, 1996.

[272] Ao Tang, Jiantao Wang, Sanjay Hegde, and Steven H. Low. Equilibrium and fairness of networks shared by tcp reno and vegas/fast. *Telecommunication Systems*, 30(4):417–439, 2005.

[273] Gang Tao. *Adaptive Control Design and Analysis (Adaptive and Learning Systems for Signal Processing, Communications and Control Series)*. John Wiley and Sons, New York, NY, 2003.

[274] Richard Thommes and Mark Coates. Deterministic packet marking for congestion price estimation. In *Proceedings of IEEE International Conference on Computer Communications Twenty-Third Annual Joint Conference of the IEEE Computer and Communications Societies*, pages 1330–1339, March 2004.

[275] Richard Thommes and Mark Coates. Deterministic packet marking for time-varying congestion price estimation. *IEEE/ACM Transactions on Networking*, 14(3):592–602, 2006.

[276] Yu-Ping Tian. A general stability criterion for congestion control with diverse communication delays. *Automatica*, 41(7):1255–1262, 2005.

[277] Tuan Anh Trinh and Sándor Molnár. A comprehensive performance analysis of random early detection mechanism. *Telecommunication Systems*, 25(1-2):9–31, 2004.

[278] Georgios Tselentis, John Domingue, Alex Galis, Anastasius Gavras, and David Hausheer. *Towards the Future Internet: A European Research Perspective*. IOS Press, Amsterdam, The Netherlands, 2009.

[279] Andrea De Vendictis, Andrea Baiocchi, and Michela Bonacci. Analysis and enhancement of tcp vegas congestion control in a mixed tcp vegas and tcp reno network scenario. *Performance Evaluation*, 53(3):225–253, 2003.

[280] Elvis Vieira and Michael Bauer. Proactively controlling round-trip time variation and packet drops using smooth TCP-q. In *Proceedings of the 3rd International Conference on Quality of Service in Heterogeneous Wired/Wireless Networks*, pages 39–48, August 2006.

[281] Glenn Vinnicombe. On the stability of networks operating TCP-like congestion control. In *Proceedings of 15th International Federation of Automatic Control World Congress*, pages 685–702, July 2002.

[282] Thomas Voice. Stability of multi-path dual congestion control algorithms. In *Proceedings of the First International Conference on Performance Evaluation Methodolgies and Tools*, pages 42–56, 2006.

[283] Thomas Voice and Gaurav Raina. Stability analysis of a max-min fair rate control protocol (RCP) in a small buffer regime. *IEEE Transactions on Automatic Control*, 54(8):1908–1913, 2009.

[284] Chonggang Wang, Yiwei Thomas Hou, Kazem Sohraby, and Yu Lin. LRED: a robust active queue management scheme based on packet loss ratio. In *Proceedings of the IEEE International Conference on Computer Communications Twenty-third Annual Joint Conference of the IEEE Computer and Communications Societies*, pages 1–12, March 2004.

[285] Jiantao Wang, Ao Tang, and Steven H. Low. Local stability of FAST TCP. In *Proceedings of the 43rd IEEE Conference on Decision and Control*, pages 1023–1028, December 2004.

[286] Jiantao Wang, David X. Wei, Joon-Young Choi, and Steven H. Low. Modelling and stability of FAST TCP. *Wireless Communications*, 143:331–356, 2007.

[287] Jianxin Wang, Liang Rong, and Yunhao Liu. Design of a stabilizing AQM controller for large-delay networks based on internal model control. *Computer Communications*, 31(10):1911–1918, 2008.

[288] Junsong Wang and Ruixi Yuan. Synthesis of PI type congestion controller for AQM router in TCP/AQM network. *Lecture Notes in Computer Science*, 6328:69–76, 2010.

[289] Li-Xin Wang. *Adaptive Fuzzy Systems and Control: Design and Stability Analysis*. Prentice-Hall, Upper Saddle River, NJ, 1994.

[290] Xiao Fan Wang, Guanrong Chen, and King-Tim Ko. A stability theorem for internet congestion control. *Systems Control Letters*, 45(2):81 – 85, 2002.

[291] Zhikui Wang and Fernando Paganini. Global stability with time-delay of a primal-dual congestion control. In *Proceedings of 42nd IEEE Conference on Decision and Control*, volume 4, pages 3671–3676, December 2003.

[292] John Watkinson. *The MPEG Handbook, Second Edition*. Focal Press, Burlington, MA, 2004.

[293] David X. Wei, Cheng Jin, Steven H. Low, and Sanjay Hegde. FAST TCP: motivation, architecture, algorithms, performance. *IEEE/ACM Transactions on Networking*, 14(6):1246–1259, 2006.

[294] Michael Welzl. *Network Congestion Control: Managing Internet Traffic*. John Wiley and Sons, New York, 2005.

[295] Paul J. Werbos. *Beyond Regression: New Tools for Prediction and Analysis in the Behavioral Sciences*. PhD thesis, Harvard University, Cambridge, MA, 1974.

[296] Debbra Wetteroth. *OSI Reference Model for Telecommunications*. McGraw-Hill, New York, 2002.

[297] David White and Donald Sorge. *Handbook of Intelligent Control: Neural, Fuzzy, and Adaptive Approaches*. IEEE Press, New York, 1994.

[298] Joerg Widmer, Robert Denda, and Martin Mauve. A survey on TCP-friendly congestion control. *IEEE Network*, 15(3):28–37, 2001.

[299] Jöerg Widmer and Mark Handley. Extending equation-based congestion control to multicast applications. In *Proceedings of the Conference on Applications, Technologies, Architectures, and Protocols for Computer Communications*, pages 275–285, October 2001.

[300] Adam Wierman and Takayuki Osogami. A unified framework for modeling TCP-Vegas, TCP-SACK, and TCP-Reno. In *Proceedings of the 11th IEEE/ACM International Symposium on Modeling, Analysis and Simulation of Computer Telecommunications Systems*, pages 269–278, October 2003.

[301] Carey Williamson. Internet traffic measurement. *IEEE Internet Computing*, 5(6):70–74, 2001.

[302] Lisong Xu, Khaled Harfoush, and Injong Rhee. Binary increase congestion control for fast, long distance networks. In *Proceedings of 13th IEEE International Conference on Computer Communications*, pages 2514–2524, March 2004.

[303] Lisong Xu and Josh Helzer. Media streaming via TFRC: an analytical study of the impact of TFRC on user-perceived media quality. *Computer Networks*, 51(17):4744–4764, 2007.

[304] Ji Hoon Yang, Seung Jung Shin, Dong Kyun Lim, and Jeong Jin Kang. Robust congestion control design for input time delayed AQM system. *Communications in Computer and Information Science*, 200:349–358, 2011.

[305] Qian Yanping, Li Qi, Lin Xiangze, and Ji Wei. A stable enhanced adaptive virtual queue management algorithm for TCP networks. In *Proceedings of IEEE International Conference on Control and Automation*, pages 360–365, May 2007.

[306] Aydin Yesildirek and Frank L. Lewis. Feedback linearization using neural networks. *Automatica*, 31(11):1659–1664, 1995.

[307] Selma Yilmaz and Ibrahim Matta. On class-based isolation of udp, short-lived and long-lived tcp flows. In *Proceedings of 9th International Workshop on Modeling, Analysis, and Simulation of Computer and Telecommunication Systems*, pages 415–422, August 2001.

[308] Qinghe Yin and Steven H. Low. Convergence of REM flow control at a single link. *IEEE Communications Letters*, 5:119–121, 2000.

[309] Qinghe Yin and Steven H. Low. On stability of REM algorithm with uniform delay. In *Proceedings of the IEEE Global Telecommunications Conference*, pages 2649–2653, November 2002.

[310] Cao Yuan, Liansheng Tan, Lachlan L. H. Andrew, Wei Zhang, and Moshe Zukerman. A generalized FAST TCP scheme. *Computer Communications*, 31(14):3242–3249, 2008.

[311] C. L. Zhang, Cheng Peng Fu, Ma-Tit Yap, Chuan Heng Foh, K. K. Wong, Chiew Tong Lau, and M. K. Lai. Dynamics comparison of TCP Veno and Reno. In *Proceedings of the IEEE Global Telecommunications Conference*, volume 3, pages 1329–1333, November 2004.

[312] Heying Zhang, Lihong Peng, Baohua Fan, Jie Jiang, and Ying Zhang. Stability of FAST TCP in single-link multi-source network. In *Proceedings of the Computer Science and Information Engineering*, pages 369–373, March 2009.

[313] Ke Zhang and Cheng Peng Fu. Dynamics analysis of TCP Veno with RED. *Computer Communications*, 30(18):3778–3786, 2007.

[314] Feng Zheng and John Nelson. An H_∞ approach to congestion control design for AQM routers supporting TCP flows in wireless access networks. *Computer Networks*, 51(6):1684–1704, 2007.

[315] Bin Zhou, Cheng Peng Fu, Dah-Ming Chiu, Chiew Tong Lau, and Lek Heng Ngoh. A simple throughput model for TCP Veno. In *Proceedings of the IEEE International Conference on Communications*, pages 5395–5400, June 2006.

[316] Bin Zhou, Cheng Peng Fu, Ke Zhang, Chiew Tong Lau, and Cheng Peng Foh. An enhancement of TCP Veno over light-load wireless networks. *IEEE Communications Letters*, 10(6):441–443, 2006.

[317] Li Zhu, Gang Cheng, and Nirwan Ansari. Local stability of random exponential marking. *IEE Proceedings Communications*, 150(5):367–700, 2003.

[318] Nafei Zhu and Jingsha He. Quantified study on the effects of path load on the similarity of network nodes based on RTT. *JCIT: Journal of Convergence Information Technology*, 6(8):284–291, 2011.

Index

Absolutely continuous function, 223, 224

Activation function, 250, 252–255, 262

Active Queue Management (AQM), 6, 77

Actuator, 1

Adaptive congestion control framework (NNRC), 113, 118, 120, 173, 204, 207

Adaptive Random Early Detection (ARED), 69

Adaptive rate controller, 114, 117, 121, 132

Adaptive system, 225

Adaptive system statement, 225

Adaptive Virtual Queue (AVQ), 6, 71, 77

Additive increase, 84

Additive Increase-Multiplicative Decrease (AIMD), 4, 57, 59, 60, 70, 92, 93, 97, 98

Additive measurement noise, 265

Advertised window (rwnd), 51

Aggregate flow, 100

Aggregate path congestion measure, 115

Ambiguity loss, 57

Application adaptation, 204

Application layer, 29, 31

Application protocol, 32

Approximant, 249, 263, 269

Approximate representation, 248

Approximation, 248
    accuracy, 248
    error, 265
    problem, 248

structure, 248, 249

Approximation based control, 9, 108

Audiovisual network, 2

Autonomous computer, 15

Available bandwidth, 16

Backbone, 16, 21

Backlog packets, 91

Backpropagation algorithm, 262

Bandwidth, 21, 34, 49

Barbălat lemma, 226

Binary Increase Congestion Control (BIC), 84

Binary search increase, 84

BLUE, 71

Bottleneck, 81

Boundedness, 223
    definitions, 231
    solution, 231

Broadcast, 17

Browser, 32

Buffer overflow, 34

Buffering delay, 114

Burstiness control, 88

Bursty traffic, 4, 17

Cable break, 22

CBR source, 210

CDMA, 3

Central server, 23

CHOKe, 71

Circuit-switched network, 16

Client-server, 43

Closed-loop, 9, 132

Cloud, 18
    interconnecting, 19

Color depth, 204